The Hudson Valley in the ICE AGE

A Geological History & Tour

The Hudson Valley in the ICE AGE
A Geological History & Tour

Robert and Johanna Titus

Published by Black Dome Press Corp.
649 Delaware Ave., Delmar, N.Y. 12054
blackdomepress.com
(518) 439-6512

First Edition Paperback 2012
Copyright © 2012 by Robert Titus and Johanna Titus

Without limiting the rights under copyright above, no part of this publication may be reproduced, stored in or introduced into a retrieval system, or transmitted, in any form, by any means (electronic, mechanical, photocopying, recording, or otherwise), without the prior written permission of the publisher of this book.

ISBN-13: 978-1-883789-72-5
ISBN-10: 1-883789-72-9

Library of Congress Control Number: 2012946964

Front cover photograph: Copyright 2009 by Pablo Caridad.

Back cover photographs: The View from Olana. Photograph by the authors (top). Reconstruction of the Cohoes mastodon. From John M. Clarke, *James Hall of Albany: Geologist and Palaeontologist* (Albany, N.Y.: 1923) (bottom).

Text photographs: By authors unless otherwise attributed.

Design: Toelke Associates, www.toelkeassociates.com

Printed in the USA

10 9 8 7 6 5 4 3 2 1

To our latest additions:
Cindy, Tyler, and Alexis Marie

Contents

Preface		viii
Acknowledgments		xii
1.	A Journey of the Mind's Eye	1
2.	The Decline and Fall of a Warm Earth	7
3.	South by Southwest	19
4.	The Wall of Manitou	29
5.	The Great Moraine	35
6.	The Hudson Canyon	41
7.	The First Retreat	47
8.	Later Retreats	51
9.	The Drowned Lands	57
10.	Lake Albany	61
11.	The Great Deltas	67
12.	April 3rd, 16,190 BP	75
13.	The Glaciers of Plattekill Clove	83
14.	Ice Age Water Park	89
15.	The Red Chasm	99
16.	The Bottom of a Lake	109
17.	Land of Lakes	115
18.	Bash Bish Falls	121

19.	The Northern Drowned Lands	125
20.	The Wetlands	129
21.	Niagara in Philmont	133
22.	Drumlins	137
23.	March of the Glaciers	141
24.	Ice Age Architecture	145
25.	Valley of the Kings	149
26.	The Post-Glacial Fractures	155
27.	Silent Earthquakes on the Hudson	159
28.	Yellow Alert?	163
29.	The Slides of Hyde Park	167
30.	Elephant's Graveyard	173
31.	Ice Age Ghosts	179
32.	Extinction	183
33.	The Dunes of Pine Bush	187
34.	Bad Day on Wall Street	193
35.	Never Again?	197
Bibliographic Essay		199
About the Authors		205
Index		206

Preface

Welcome to the Ice Age.

This journey to a time called the Pleistocene takes you on a tour of the Hudson Valley unlike any you have taken before—you will see it as it was during the time of the glaciers. And that is not hyperbole; we mean it quite literally. Our goal is to have you see it as vividly as geologists do. Ice age history is fascinating stuff, a good subject for a book, and this is a good time to pursue it. The Hudson Valley has had a grand cultural history. We started writing this book during the quadricentennial commemoration of Henry Hudson's voyage up the river. That voyage began the recorded history of the valley. But the history of the Hudson Valley extends far beyond that, far back into the realm of the prehistoric.

For many thousands of years before Henry Hudson's arrival, this was the land of the Native Americans. But we want to take you back even farther, to a time long before even those first people arrived. The great event that heralded their migration into the Hudson Valley was the melting of the glaciers. The warming climate allowed for the reforestation of a bleak ice age terrain, and with that warming came the return of game animals and the people who hunted them.

Most people are aware that there was such a time as the Ice Age, but generally they know few of the details. Many of us have seen the feature-length animated films named under the rubric of "Ice Age." Most of us know that there were mammoths and other "ice age animals." Sadly, however, movies commonly misinform us—especially when they portray dinosaurs as living side by side with ice age mammals. Movies are fun to watch, but they are not very good sources of information about this remarkable time. We think that the science of geology can be a lot more interesting, and even more fun, than fictional representations that have a tenuous connection with facts.

We have spent years exploring the ice age history of the Hudson Valley, and we have found the reality to be everything that the movies promised ... and a great deal more. Now we invite you to join us—not just in the pages of this book, but out in the field where the Ice Age can be seen. Much of this book was written as a field guide. Pack our book in your car and go out to see the incredibly rich ice age history of the Hudson Valley. Take the family on a journey through time. There is so much to see; ice age features are everywhere, up and down the entire valley. You need to develop an eye for them, of course, but that is not all that hard to do.

You can follow in the wake of the Hudson Valley glacier as it ground its way south down the valley and then melted its way back to the north. You can follow several sizeable splinters of the ice that veered off and invaded the Catskills. Anyone knowledgeable of what to look for can follow a glacier like an animal tracker can follow a game animal. You can learn to read the signs; we are here to help.

Visit towns in the Hudson Valley that you have seen a hundred times before, and now view them through our eyes. Hyde Park and Schenectady are among many locations that originally formed as deltas along the shores of a great ice age lake called Glacial Lake Albany. Albany, not surprisingly, lies at what was the bottom of that enormous lake. Palenville is a heap of sand, piled up at the bottom of a great pounding ice age stream, a thunderous version of modern Kaaterskill Creek. Greenville is shaped like the bottom of a glacier, and for good reason—it was shaped by the moving and advancing ice.

Some of our region's most scenic landscapes are the results of the glaciers. All of the Shawangunk Mountains were sculpted by the ice. So, too, was the Catskill Front. When you visit North/South Lake State Park, you are returning to the Ice Age.

Our landscapes are renowned for the art they inspired. The Hudson River School of landscape painting was founded here. Much, perhaps most, of what those artists portrayed was glacial in its origins. They would have had a lot less to inspire them were it not for the Ice Age.

The Hudson Valley also played a very important role in the development of America's landscape architecture, with some of its most important pioneers practicing here. They built upon a landscape that the glaciers had shaped first. We believe that the ice influenced our country's ability to develop its remarkable landscape architecture.

In short, all of the Hudson Valley can be considered a gift of the Ice Age. Its residents should be aware of that and should have an appreciation of the prehistory that surrounds them everywhere and every day. That is our purpose.

A few caveats are in order. The ice age record of the Hudson Valley is terribly complex. The evidence is extraordinarily extensive and often difficult to interpret in all its details. Ironically, a rich prehistoric record can cloud a person's view of the past. We have not hesitated to generalize what happened for the purpose of illuminating and clarifying those events. Our efforts are to paint a broad picture of the forest; we will not describe many of the trees.

The professional literature is written by scientists and for scientists; it is commonly difficult to understand. We are experienced scientists, but we are not writing like scientists here; we are not writing this book as something to send in for peer review. We have written this account for the general reading public. Much of this book is autobiographical. We describe our adventures with the Ice Age as we lived them. The purpose is that we sincerely wish that our readers can explore this land as we have and see it as we do.

We frequently use literary devices that we would never use in the scientific canon. We are fond of fantasies, especially in chapters one and twelve. We will commonly endeavor to portray the past as if we had been there observing events as they happened. Such fantasies are "informed portrayals," not science. Our fantasies are re-creations, not documented facts. We ask our readers and colleagues to have the good judgment to understand this and not to take such passages overly seriously. This is artistic license, and its purpose is to communicate the essence of the past, perhaps even the romance of the past, but it should not be considered as literal or factual representation of the past.

We do describe some ice age geology that we do not think has been adequately recorded in the professional literature. For example, we document an abundance of ice age hills, called drumlins, in more detail than we have found in the scientific literature, and our descriptions of glacial spillways at North Lake have scarcely been covered in previous publications. But we wish to emphatically disclaim any professional intentions with this book. We stake no professional "claims" to the Hudson Valley Ice Age. We solely aim to illuminate the general public and to awaken interest in a remarkable place and time.

For these reasons we will list relatively few references to the professional literature—limited numbers of the more important and general works that have been done. This book is popular literature for a general readership.

And that is a very worthwhile goal. The ice age history of our Hudson Valley is an epic story. We describe the journey of glacial ice, thousands of feet thick, moving inexorably down our valley. We watch as it grinds its way into the landscape beneath. We travel forward through time and watch as a changing climate begins to melt this huge mass of ice. We then witness vast quantities of meltwater flow across the landscape, forming large lakes and gouging out great channels. All this imagery makes the Ice Age a wonderful topic for popular science writing. And all this is just plain worth knowing.

Enjoy your journey through the Hudson Valley and the Ice Age, a strange and wonderful journey through a landscape you thought you knew.

Robert Titus
Johanna Titus
Freehold, New York
July 2012

Acknowledgments

This book is the product of twenty years of popular science writing. We have been writing for *Kaatskill Life* magazine, the *Woodstock Times*, the Columbia County *Independent* and the *Register Star* newspaper chain. Meeting deadlines hones your skills at writing, and writing for the general public sharpens your skills at communicating. We think that shows in this book. Our thanks, foremost, go to *Kaatskill Life* for taking a chance on adding a geology column to its pages. It was a chancy decision for them, and that choice transformed our lives.

The fieldwork for this book was extensive and was largely funded by the winning of a Win Wanderslee Award from Hartwick College. Win Wanderslee was a Hartwick professor who left funding for the award in her will. We thank her personally.

The staffs at the great homes of the Hudson Valley, particularly at Olana, Wilderstein, and Springwood, were very welcoming and helpful; so, too, were the staffs at Pine Bush Preserve.

The New York State Museum and Woods Hole Oceanographic Institute were generous in letting us use some of their graphics.

We learned a great deal in the field by joining the hikes of the Mountain Top Historical Society. Hike leader Bob Gildersleeve was a knowledgeable guide who is well versed on our region's geological history. He was very helpful.

Our thanks to Ron Toelke Associates for the fine job they did of laying out the design of this book.

And thanks to our families and friends, and especially our son Chris Offner and granddaughter Azia Layman for helping us to flesh out the visual images of our "mind's eye."

A Journey of the Mind's Eye

January 15th, 22,090 BP.[1] We are the mind's eye, the human imagination, and we are gliding through the air, sixty miles or so in altitude above the vicinity that will lie, in our own modern times, southeast of the great harbor that is at the mouth of the Hudson River. We gaze forward and, strangely, we see no such river. To our right should be Long Island; to our left should be the northern New Jersey coast with its familiar Highlands. But not one bit of any of this can be seen.

We are confused—more by what we don't see, than by what we do. Below, there is no Atlantic Ocean. Instead of an ocean, there is a flat landscape that is about as barren as can be imagined. We can observe only one thing from up here—the evidence of a substantial stream. We descend to look more carefully, and we do see the multiple crisscrossing channels of a complex river, but there is no water in any of them. They comprise a dry version of what geologists call a **braided stream** (fig. 1-1), and they speak of moments in time when large volumes of water flowed out onto this barren land surface and swept along enormous amounts of sand. The rush of water created those braided channels, and each was, for a while, glutted with all that sediment. But where did all this come from?

Figure 1-1. Braided streams. Courtesy U.S. Geological Survey.

Why had this strange stream been here? And where did it go?

Now our eyes are a bit more trained and we see that there is a lot of sand down there, and it is all over the place. We see numerous sand dunes scattered about. We realize that there must be times when it is very windy around here.

We are the mind's eye; we can go anywhere and we can do anything. Now we drop down even lower and pass across and just above the ground. We see that nature has been struggling to establish a ground cover of foliage on this bleak and sandy landscape. But, as of now, there is only a patchy flora here composed of sedges, grasses, mosses, and lichens—plants that in modern times would make up the habitat we call **tundra**.

What mystery is this? What land are we visiting? We should be offshore of what will someday be the metropolitan New York region. It will be one of the world's greatest port cities. How can there be barren tundra here instead of an ocean? We rise up into the sky and continue forward, using the path of those braided stream channels as our compass. Now, far to our southwest, we see what seems to be a more established ecology. Out there is a forest of spruce and pine. These are the Pine Barrens of the southernmost New Jersey coastal plain, and at least now there is something familiar. Turning, we can also see a similar well-established landscape off to our northeast. This is Connecticut, and there are sizable forests out there, too. At least that is what they look like from this high up and this far away; we no longer think that we can be sure about anything!

But a few familiar landscapes don't help much to relieve our growing anxieties. These mysteries are deep and puzzling. Where is the Atlantic Ocean? What has happened to it? And where, for that matter, is the Hudson River? Dry braided stream channels are no substitute. The river should be visible somewhere nearby. But there is something else missing; there is still no sign of anything resembling Long Island. It is as if we had entered into an alien landscape. All this is just too strange, too disorienting. We continue forward.

Now we make our final approach to what should be New York City, and our bewilderment only gets worse. We do see something that resembles Manhattan Island. However—and not surprisingly—there is no city here. This "Manhattan" is surrounded by the channels of what should be the Hudson and East Rivers, but these channels are empty and dry. To the west, it gets better; there are the familiar cliffs of the Palisades. Here, once again at least, is something that looks right!

We drop down and drift slowly above the "Hudson River." Now we see several other empty, braided stream channels. A pattern has been established, and we can follow some hunches. Periodically, it would seem, the Hudson River "wakes up" and is flushed with large volumes of water that flow for only limited periods of time, sweeping along great quantities of sediment. Then the river goes dormant once again. The water drains away, but those dry braided channels are left behind. We feel reassured that we are able to solve at least one small mystery. That helps—a little—but that still really doesn't explain anything important; we have, in fact, just added another mystery. What is turning the river on and off? What can explain all of these mysteries? Now we are not just disoriented; we are awed. We continue north.

We soon reach what should be the Tappan Zee. It is, on this day, a breathtaking sight—a great, dry, empty hollow, miles across and many more miles long. Braided stream channels are seen at its bottom. The slopes of the Tappan Zee basin display old, and now dry, shorelines. There are many of them, and they combine to form a banding within the basin. Evidently, again and again it has been filled with water and emptied. Each episode created a temporary shoreline, and each shoreline left its own mark—the banding that we see. What made such things happen?

The mysteries abound. Along the ridges above the Tappan Zee can be seen a forest in the dead of winter. From our far distance, it just does not look right. Deep snows bury all of the lower portions of the tree trunks. It would seem that we have arrived in the middle of a very bad, even terrible, winter season, but there must be something else. We look again, but we just cannot tell what that might be. We are compelled to see more; we are driven onward by that compulsion. We continue north.

We are drifting faster now, continuing up the Hudson Valley, and there is no mistaking the geography. Other than the fact that it continues to lack water, it does look like the landscape we know. We can recognize landmarks in the valley. West Point is easy; it hasn't changed much, except that it sits at the edge of a dry channel. The same can be said for Storm King. We convince ourselves that we can recognize where Poughkeepsie is, even though we really are not sure. Then, off to the west, are the Shawangunk Mountains. Next, the low ridges near Kingston tell us that we have arrived near that city. We drop down again to a very low altitude. Now we encounter a new mystery. We welcome this one even less than all those that came before it.

We are now close enough to the forests to finally perceive that peculiar something that we had missed before, but which had unsettled us. It is winter, and all of the trees look dead, but now we see that most, and probably all, of them *are* dead. All of them have been stripped of their twigs, all of them have lost branches, and some of them have even lost large, thick limbs. This forest has been splintered! There is no doubting it; all of the trees along the Hudson have died. There must have been ferocious winds that ripped off much of their dead and desiccated branching. Dead forests! All our mysteries are now compounding. We are not just disoriented and awed; we are frightened. We are horrified. We are now panicked. But ... we continue on.

Now we rise up fifty miles into the sky; then we rise another fifty. We turn around and look behind, back even beyond where we began our odyssey. There, in that far distance, at last, we see the Atlantic Ocean. It has retreated a hundred miles or so from the shoreline that we have known in our own times. Sea levels have dropped, and that great ocean has drained away. That partially explains the windswept tundra.

The retreat of the sea is still one more perplexing mystery, but not for long. We turn around again and look forward into the north and— our mysteries are suddenly, instantly, all solved! Out there, far to the north, the Hudson Valley is filled with ice—a glacier. But beyond that lies the broad white of something much greater. It's called an **ice sheet**. It spreads out, west to east, across the whole northern horizon rising above the Hudson Valley. A great winter has been descending this valley and bringing with it this mass of ice called the **Laurentide ice sheet**. And now, in effect, our look into the north is also a look into the future; this ice sheet is advancing and it will, given time, overrun all the rest of the Hudson Valley.

The vastness of this glacier gives it a long reach. Its expanding mass has actually been robbing the oceans of their waters. Such waters are always evaporating, but now their vapors are falling as snow onto the ice, making the ice larger. As the ice sheet grows and advances, more water leaves the sea and becomes trapped in even more glacial ice. There is none of the usual return of the water to the oceans via rivers; the rivers, as we have seen, have been drying up. The sea levels are dropping, and those shorelines are retreating. The seafloors are being exposed. And they are exposed to a very cold climate.

It is this expanding glacier that has robbed the Hudson Valley of its waters. There are occasional reversals, short periods when some warmth returns, and that is when the Hudson "wakes up" and produces those braided streams, but those moments are becoming increasingly rare.

We, the mind's eye, are dazed as we gaze at these wonders. Our lips form the silent words, "How did all this happen?"

---

[1] For those unfamiliar with BP dates, BP stands for "before present," with "present" being 1950, the year carbon-14 calibration curves were established. It is the now-accepted convention for applying dates to prehistoric events.

2

The Decline and Fall of a Warm Earth

It is difficult to determine the origins of any major geological event. What causes an earthquake to begin at any particular moment? What influences result in the eruption of a volcano? Both are sudden and dramatic geological events; they strike and they kill and then they are over. They may seem to occur spontaneously, but they are actually the products of a prolonged sequence of earlier events—earlier stresses that had built up slowly over enormous lengths of time. Thus it is that geologic processes reach forward through time, often through many millions of years, to generate future events, unforeseen events. Right now, at this moment, prospective earthquakes and potential eruptions are developing ... somewhere.

If all the geological processes going on today have their origins in the distant past, then so it must have been with the Ice Age, too. There is nothing sudden about an ice age; it builds slowly over millions of years. When it strikes, it strikes slowly; it kills, but again, slowly, over immense lengths of time. What brings on an ice age? That is the subject of this chapter. This is the hard chapter, because it is all scientific theory, so it is perhaps good to get it out of the way early.

Ice ages are exceptional events. Throughout much of Earth's history, there were long periods when it was quite warm and there was very little or no ice on our planet. That's probably the norm for Earth. Some stretches of especially tropical climate are often referred to as episodes of a **Greenhouse Earth**. A good example is the early Carboniferous Period, about 360 to 320 million years ago. It was a time of beautiful Bahamas-like tropical seas all around the globe. The late Cretaceous Period, roughly 100 to 65 million years ago, was another good episode of a Greenhouse Earth chapter. It is a time also remembered for its vast tropical seas with great coral reefs that were widespread all across the globe. A third and most recent of these Greenhouse Earth times occurred during the early Eocene Epoch, about 55 million years ago, when the world's climate may have been 15 degrees Celsius warmer than it is today. This was an era of great volcanic eruptions along the mid-ocean ridges. Greenhouse gases—methane and carbon dioxide—belched forth into the atmosphere. It was, for a relatively substantial time, a very warm world.

Greenhouse gases allow sunlight to easily enter into the atmosphere. Some of the high-energy solar radiation reaching the Earth's surface is reflected back into space as lower-energy infrared waves, or more simply, as heat energy. But, when greenhouse gasses build up in the atmosphere, they absorb this heat before it can escape into space. That heat is trapped in the same manner that the glass of a greenhouse allows the sun's radiation in, but traps the lower-energy radiation (the heat) that would otherwise be reflected back out. A greenhouse can be quite warm on a sunny, but otherwise cold, winter's day. Likewise, the Earth's climate warms up dramatically whenever those heat-trapping greenhouse gases are abundant; they make up the "glass" of a greenhouse Earth.

The early Eocene seems to have witnessed this, and may have been as warm a time as there ever has been on Earth. There was, for a short period, a balmy tropical world, completely ice-free even in Antarctica. It wouldn't last; an ice age was coming.

Just as there have been long periods of Greenhouse Earth conditions, there have also been times of **Icehouse Earth**. These most likely occurred when greenhouse gases were relatively scarce. Icehouse Earth intervals have occurred several times during the last half-billion years. Such episodes of Icehouse Earth conditions can descend into full-blown ice ages. These can persist, off and on, for millions of years or even longer, and are defined as times when continental ice sheets and glaciers are present across much of the Earth's surface. That's what happened during the **Pleistocene**, the most recent three million years.

During an ice age there can be **glacial epochs**, periods of great cold when ice sheets advance and cover wide expanses of the Earth. There can also be times of relative warmth, **interglacial periods**, when the great ice sheets melt back but do not completely disappear. We are living in the midst of such an interglacial period. The late Pleistocene time we commonly refer to as "*the* Ice Age" is simply the most recent of perhaps more than twenty or so episodes of glacial advance and retreat (epochs) making up the whole of the present Ice Age. The last glacial advance, which peaked about 21,000 years ago, is known as the **Wisconsin Stage**. That one is the topic of our book. When we use the term Ice Age, we are almost always speaking of this particular time.

Geologists for decades have speculated about what caused the Ice Age. Textbooks vary considerably in their explanations. We will offer an account in this chapter, but must emphasize that all of this is still quite tentative. How, when, and where are the important questions. *How* did it all begin? There are many processes that can contribute to global cooling, processes that together could cause an ice age to begin. Over tens of millions of years, patterns of tectonic plate movements, called **Wilson cycles**, cause the redistribution of great land masses and oceans. Throughout these cycles continents collide, and mountain ranges are uplifted and then eroded away, sometimes only to rise again. Meanwhile, ocean basins form and reform. All these processes, and more, contribute to great changes in the global climate.

When and *where* did the Wilson cycles begin to lead our planet into an ice age? This is very debatable, but one possible answer is 37 million years ago in the southern hemisphere. There must have been a day when a small fissure first opened up between Antarctica and Australia, and it gradually grew into a great geological fault. The two landmasses had been locked in a continental embrace for vast lengths of time, but at about 37 million years ago they began a split; Australia drifted north, leaving Antarctica still lying upon the South Pole. Changes in ocean currents and new climate patterns resulted, and these gradually cooled the south polar continent. South America, too, had been attached to Antarctica and, later in time, began a separation that further isolated Antarctica. With that separation, a large circumpolar current of sea water formed and rotated around the now increasingly remote southern continent (fig. 2-1). That's important, as it created a great wall of water that further cut off Antarctica climatically and shut it off from all outside sources of oceanic warmth. Antarctica, now located by itself astride the South Pole, gradually cooled. Glaciers formed and began to spread across its polar landscape, possibly about 34 million years ago. It was a harbinger of the Ice Age. But it was just the beginning; it would get worse.

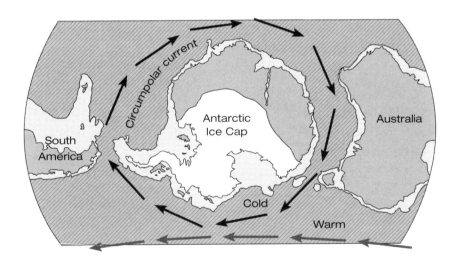

Figure 2-1. Map of Antarctica with Australia and South America backing away. Note the circumpolar current in the seas and the Roaring 40's wind patterns. Modified from Prothero and Dott, 2011.

During these times an unstable Earth would produce another crustal event. India had once been a subcontinent attached, not to Asia, but to east Africa. It had detached and begun a slow journey across the Indian Ocean. At roughly 36 million years ago, it began a collision with Asia that still continues today. That collision gradually induced the uplift of the world's greatest mountain range, the Himalayas. Rising mountain ranges bring enormous volumes of rock into contact with the vicissitudes of climate. The result is that these towering mountain ranges gradually began to weather away.

The many processes of chemical weathering include one in particular—carbon dioxide attacks rock and breaks it down. The carbon dioxide itself is destroyed in the process so, as a side effect, the world's atmosphere loses significant amounts of its most important greenhouse gas. As carbon dioxide is lost, the greenhouse gas blanket over the Earth thins, and heat is lost; little CO_2 is left to stop it as it radiates off into space. This helps the world to make its transition into an Icehouse Earth episode.

But we don't want to imply that there is anything simple about any of these Wilson cycle patterns. Over the last 30 million years there has also been uplift of great mountain ranges throughout western North America and western South America that has equally promoted an Icehouse Earth effect. The rising of these great landscapes has had a significant effect on climate patterns, diverting winds and causing changes in monsoonal wind circulation. The great oceanic currents were altered as well. That's pretty complex stuff, with equally complex effects on climate. Mostly it made the Earth colder. But, it would get worse.

At about 23 million years ago, during a time called the Miocene Epoch, South America completed its separation from Antarctica, and the Drake Passage between them widened considerably. A great circumpolar climate pattern, a wind current called the "Roaring 40's," enveloped Antarctica. As South America backed farther away from Antarctica, this circumpolar air current was greatly enhanced. Its fierce winds made worse the isolation of Antarctica that had begun earlier (fig. 2-1). The great southern continent now lay within a prison, walled off by both water currents and wind patterns. Very little warm air or water could get in, and not much cold air or water could escape. Gradually the entire Antarctic land mass slipped deeper into its own private ice age. Now bigger, thicker, and colder glaciers would advance across its landscape, and Antarctica would become as we know it today. It would get worse.

About 14 million years ago the action shifted northward. The Iceland–Faeroe Island ridge, stretching across the North Atlantic, sank. The ridge had long been a dam, holding very cold water within the Arctic Sea and keeping it out of the North Atlantic. Now those cold waters "escaped"; they rushed out into the North Atlantic, widening and deepening the cold realms of a now rapidly cooling Earth. As a result, moist cold air flowed southward, allowing even more ice to build up and affecting land as far away as in Antarctica. There, great ice sheets formed and began to flow toward and into the surrounding seas. Now the pace of events was picking up.

Next, at about six million years ago, the glaciers of Antarctica began another major expansion, and shelf ice flowed out into the southern oceans. So much ice formed that sea levels, all over the world, dropped dramatically. This was one of the last steps; the Ice Age was now very serious business.

But, it would still get worse.

Three and a half million years ago, the Isthmus of Panama was especially active. This actively volcanic belt rose out of the sea and formed a link joining North and South America; the two had previously been separate. This blocked the flow of water from the Atlantic to the Pacific, and diverted that current to the northeast. That led to the birth of our modern Gulf Stream. Warm water currents, accompanied by humid air masses, could no longer flow into the Pacific. They turned around and started flowing northeasterly instead, across the Atlantic. By the time these currents reached Europe they had cooled down. Cold water is denser than warm water, and when the currents became dense enough, just north of Iceland, they sank deep into the Atlantic and began returning southward, at great depths. Thus, very little warm water reached the

Atlantic's Polar regions. But the humid and moist air currents that had accompanied the Gulf Stream did continue onward into the north, and they began to actively produce snow there. Wherever there was land, the snow accumulated. The snow piled up in Labrador, Greenland, and in Scandinavia.

With increased precipitation in these regions, the flow of rivers into the Arctic Sea increased, and that flow caused the ocean to become a little less salty. This made it easier for the ocean's surface water to freeze, and now large stretches did become covered with ice. This was the gradual beginning of the Pleistocene glaciation of North America. Incredibly, it would get worse.

As the Arctic and North Atlantic realms cooled, more snow fell than melted. The snow compacted into ice and, when it was thick enough, it began to flow in the form of glaciers that expanded outward from their centers of accumulation. There were four of these centers. One formed in western Canada—the Keewatin Center. A second center formed over Scandinavia. The third blanketed Greenland. The fourth and, from our point of view, the most important, formed in Labrador and flowed out to all of the northeastern United States, including the Hudson Valley (fig. 2-2).

Snow is white and reflects more sunlight back into space than the darker, more absorptive lands not covered by snow. For the same reason, we wear white in the heat of the summer and darker clothes in winter. Colors and shades have different reflective powers. We call that effect on the Earth its **albedo**. Each of those centers of accumulation collected more and more snow, and that new snow was a brilliant white; thus it reflected still more sunlight, producing a much higher albedo. Much of the sunlight that was reflected ended up radiating back into space. The Earth was now actually losing its solar heat altogether! Events had turned a corner. The snowy northern hemisphere was actually cooling itself. Things had gotten worse!

The events that we have been chronicling so far have all described an Earth that was falling into colder times. Rising continents had deprived it of its major greenhouse gas, CO_2. The isolation of Antarctica had robbed that land of all sources of heat. The Gulf Stream had generated abundant snowfall in the northern latitudes. The Earth's albedo was rising. All these trends seem to have been dragging the Earth into a colder time, and that might seem enough to account for the onset of the Ice Age. But it wasn't.

Our story has, so far, followed the Earth through tens of millions of years to a point when it was just cold enough so that glaciation might be expected. But we need something else; our theory is not yet complete. Ice ages occur in roughly 100,000-year cycles, and Wilson cycles are much too long to possibly cause anything like this. We still have to

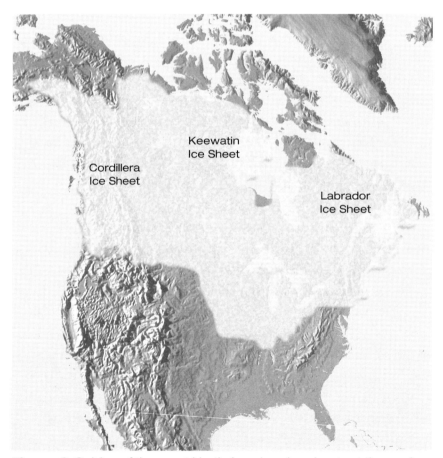

Figure 2-2. Map of the great North American ice sheets at the peak of the Ice Age. Courtesy U.S. Geological Survey.

account for how it is that the Earth has wandered into and out of these relatively short 100,000-year cycles of glaciation. Many geologists think the answer lies in relatively small differences in the orbit of the Earth around the sun. These, it can be argued, could allow the ice to accumulate, form glaciers, and advance in cycles of glaciation. In short, we shift from geography to astronomy.

The changing tilt of the Earth and its elliptical orbit around the sun are the causes of our planet's seasonality. If the Earth's orbits were circular instead of elliptical, and its axis was unchanging, there would be no seasons; any given point on the Earth would always receive the same amount of solar **insolation** (almost the same as sunlight) all throughout the year and all throughout time. It's not like that, however; there are, instead, major astronomical cycles. Over the past generation or so, scientists have gravitated toward an explanation for glaciations that involves these astronomical cycles. They are called **Milankovitch cycles**, named after their Yugoslav discoverer, Milutin Milankovitch. These cycles describe the long-term changes in the Earth's tilt and orbit around the sun over periods of time that reach tens and hundreds of thousands of years.

As Milankovitch cycles peak and ebb, they cause variations in the amount of solar insolation reaching the Earth's surface. These changes in insolation—especially at 65 degrees north latitude, where great ice sheets form in the northern hemisphere—have been found to be important contributors to the cycles in the growth and retreat of ice sheets.

The most important type of Milankovitch cycle is Earth's **eccentricity**. At regular, long intervals (about 100,000 years), the pull of gravity from the massive planets Saturn and Jupiter actually stretches the Earth's orbit, causing it to be farther from the sun at one time of the year than it is exactly six months later. At other times, when that gravitational pull is at a minimum, the opposite occurs and the Earth's orbit is nearly circular, showing only very little eccentricity. At maximum eccentricity the Earth can receive 20 to 30 percent less insolation during the season when it is farthest from the sun. The reverse (warming) occurs with minimum eccentricity (fig. 2-3).

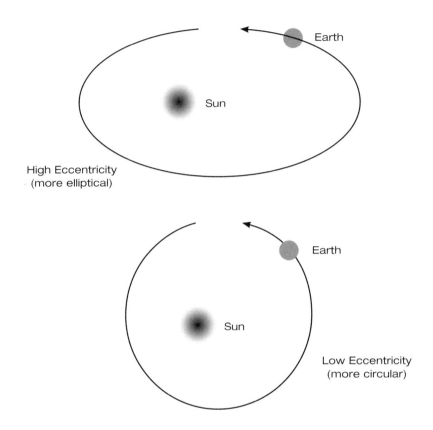

Figure 2-3. An eccentric Earth orbit around the sun. Modified from Prothero and Dott, 2011.

Here's the important part: if winter occurs at a time when the Earth's orbit is eccentric and at a point where it is closest to the sun—at perihelion—then the northern hemisphere experiences relatively warm, moist winters. When summer occurs, six months later, its orbit is at its maximum distance from the sun—at aphelion—causing cool summers. The result of all this is a lot of precipitation (snow) in the warm moist winters, but relatively little melting during the cool summers. Many argue that it is the lack of summer melting that is most important.

Along with eccentricity cycles, the Earth's angles of axial tilt, known as **obliquity**, and **precession** play a role in the insolation reaching

Earth's surface. Earth's axis, the line through the north and the south poles, is not perpendicular to its orbit around the sun. It orbits the sun at an angle. That angle, the axial tilt, varies by about three degrees over a period of every 40,000 years. At its smallest degree of tilt, winters in the northern hemisphere are warmer, and summers in the southern hemisphere are cooler, also promoting more northern snow and less southern melting.

Precession describes the Earth's wobble as it spins on its axis. When this wobble causes one hemisphere or the other to point more toward the sun at the point in the Earth's orbit when it is closest to the sun, that hemisphere will have warmer summers but colder winters. The opposite hemisphere will have cooler summers and warmer winters.

Here's the most important part. *These three cycles all contribute to the amount of solar insolation reaching the Earth's surface, and when they act together to cause warmer winters and cooler summers, especially at 65 degrees north latitude, they induce the growth of ice sheets—periods of glaciation.*

When they act to produce warmer summers and cooler winters in the north, then it is too cold and dry in winter for much snowfall. The following summer sees the Earth close to the sun and so warm that any leftover snow will be melted away quickly. The result is an interglacial time.

At present, the Earth's orbital eccentricity is nearly at a minimum in its cycle, the axial tilt is in the middle of its range, and precession has the northern hemisphere pointing away from the sun in winter. What does all this mean? You can expect another ice age in about 50,000 years … maybe.

Generally, our account of the onset of the Ice Age requires that the Earth over the past 50 million years sank into colder times, mostly for the geographic causes related to the Wilson cycles that we discussed earlier. That cooling brought our planet right to the brink of ice age cold. Having achieved this, the astronomical cycles took over and have been cycling us into and out of ice ages ever since.

So … *when*, exactly, was it that *the* Ice Age began? Well, you see

... it's a long story. How did it happen? Well, it might be fairest to say that a large number of factors combined to *conspire* to bring us into the Ice Age.

3

South by Southwest

To live in a landscape that is being approached by a large advancing glacier must be an awesome and frightening experience. Our ice age ancestors in Europe certainly lived at such a time and in such a land. What must they have thought of it all? What did the shamans tell them? We can't know; we can only speculate.

As scientists we can do a lot more than just speculate. We can assemble the evidence and put it all together to form an image of what the old landscapes must have been like. But, what kind of evidence can we search for? The problem with ice ages is that they are very destructive. During any ice age, glaciers advance and retreat many times. Advancing glaciers destroy the evidence of … advancing glaciers! They bulldoze their ways across the landscape and destroy most of the clues that might have told us what had been there earlier. Each cycle of advance and retreat destroys most of the evidence from any other, earlier glaciation. During continued advance or later retreat it may also destroy any evidence that it had itself left just a little earlier.

Fortunately, there are some regions in the Hudson Valley where the evidence of advancing glaciers is still abundant and well preserved. One of

Figure 3-1. The "Sleeping Giant" in the far distance—a view from Sunset Rock.

those locations lies within the Shawangunk Mountains. From most places in the Hudson Valley, you can spot near New Paltz the image of what is called the "Sleeping Giant." That's the profile of "the Gunks" (fig. 3-1).

Much of the Gunks has been set aside as parkland or nature preserve, and that makes them a fine place for visits. Within this area lies Minnewaska State Park, one among many of the Hudson Valley's picturesque locations. The park's scenic centerpiece is at Millbrook Mountain, nestled on a high plateau of brilliantly white sandstone. All across Minnewaska Park are numerous carriage trails that mostly date back to the nineteenth century. There are also hiking trails that take you out to the edge of the plateau for grand views of the Hudson Valley and up and down the whole of the Shawangunk Ridge. Although carriage rides are no longer available in this day and age, you can still swim and hike and picnic. For ice age geologists, the main recreation is chasing glaciers.

Between roughly 40,000 and 18,000 years ago the area experienced several episodes of glaciation. The big one was about 21,000 years ago when the Laurentide ice sheet swept out of Labrador and down across much of North America, including Connecticut, the Catskills and the entire Hudson Valley. While the leading edge of the glacier's advance was probably confined to the Hudson Valley (we call this the **Hudson Valley Glacier**), it eventually became large enough and thick enough to be able to flow over the entire Shawangunk Ridge and then onto and over the Catskills as its leading edge advanced slowly to the south. The ice sheet that followed must have reached several thousand feet in thickness, at the least.

Not surprisingly, this had a great effect upon the hilltops and slopes of the Shawangunks. In fact, the Gunks are among the best places to go and investigate the advance of the ice and to truly appreciate that advance.

You won't have to spend much time looking for evidence of glaciation at Minnewaska. It's waiting for you right there at the parking lot. Get out of your car and you will see a number of rock outcroppings that have been overridden by the ice (fig. 3-2). As the ice slowly advanced across the bedrock, it scoured and sanded the rock, giving it and much of the Shawangunk Ridge a smoothed, even polished, appearance. Some of the

Figure 3-2. Parking lot outcrops at Lake Minnewaska. Glaciers moved right to left.

Figure 3-3. Shiny rocks with striations on the Blue Trail at Lake Minnewaska.

exposures can actually gleam in the morning sunlight (fig. 3-3). That may surprise you, as you might not expect glaciers could polish rocks quite so well. But there are a lot of shiny rocks at Minnewaska.

Sanding and polishing are just the right words for what has happened here; the advancing ice carried with it a great deal of sand, and that sand was concentrated at the bottom of the ice. Thus the glacier behaved exactly like sandpaper, and we can clearly see the results. Cobbles often were dragged along as well and, as these scraped the bedrock, they left many long, straight scratches that we call **striations** (lower part of fig. 3-4).

There are also some prominent crescent-shaped chips in the bedrock; they are called **chattermarks** or **crescent marks**. They are nested, which means that the crescents are all lined up and parallel to each other (middle of fig. 3-4). These are the marks of boulders caught up in the advancing glaciers. Boulders cannot always be dragged along smoothly at the bottom of the ice. Instead they advance in periodic "jumps." The relentless push of the advancing glacier tries to press the

Figure 3-4. Striations at bottom of picture; crescents or chattermarks in the middle.

boulder forward. Meanwhile the weight of the ice presses it down. At the same time, friction tries to hold the boulder in place. There is, as you can see, a lot of stress at the bottom of the ice, all focused upon the poor boulders. Eventually it is the push from behind that "wins." When the power of that forward shove exceeds the frictional drag, the boulder leaps ahead quickly and pounds the underlying bedrock at its place of "landing." That impact forms the crescent fracture. The process is soon repeated and, with time, a linear sequence of crescent marks is produced on the rock surface. Polished surfaces, striations, and linear crescent marks all make very eye-catching sights. Once your eyes are trained to watch for them, such glaciated landscapes are very easy to recognize. They are common, and they are important. They are the marks of an advancing glacier.

Those first outcrops at the parking lot set the tone for a hike at Minnewaska. Take the red-marked Millbrook Drive. It takes you past the lake and south toward the escarpment. It's a rough hike along some stretches, but worth the effort. Take a compass with you, and you can work out a little glacial history along the way. The striations are important. Each one has a compass direction; they range between 20 and 40 degrees west of south. That southwest movement records the path of the advancing ice, and that direction is typical of the whole region.

What is most surprising, however, is how abundant those striations and crescents are. They are everywhere. It's the same with glacial polishing. Maybe it shouldn't be a surprise, though, as the Shawangunks are largely composed of quartz sandstone. That's a sturdy type of rock, highly resistant to weathering and erosion. It is thus very hard to weather the glacial features away.

Continue south on the red trail and begin an ascent to the edge of the escarpment. Along the way there are many more broad sanded surfaces. The glaciers had been very active here and had really polished the outcroppings. Along the trail the slope becomes quite steep and the soils are very thin; many patches of polished and striated bedrock are visible, peeking through holes in the mountain soils.

It didn't much surprise us that the soils were thin. What did surprise us was that there were any soils at all. Why wasn't all the landscape just pure bare rock? After all, a great glacier had crossed this way. Why hadn't it bulldozed away all of the soils? The answer is that the soils left here make up what we call a **ground moraine**. These are thin veneers of glacial sediment left behind by the slowly melting ice at the end of local glaciation. The glacier had been "dirty" with sediment, and plenty of such sediment is still here. There's always lots of ground moraine left in low-lying landscapes, but it thins out quite a bit toward the tops of the hills.

At the top of Millbrook Mountain, you come out onto the great escarpment that marks the southeastern flank of the Gunks. This cliff of quartz sandstone offers a wonderful vista of the middle Hudson Valley.

You can't resist picking a comfortable rock and sitting down to enjoy the view. Up here, however, there are other distractions; there is a lot more glacial history to be seen. You can see where the passing ice had ripped loose great masses of rock. The process is called **plucking**. Passing ice adheres to the bedrock and yanks upon it. Typically, bedrock has many fractures within it (fig. 3-5). These serve as lines of weakness and, as the ice moves along, it engulfs large blocks, loosens them, and drags them off. As this process continued throughout all of the Ice Age, it did much to shape the steep front of the escarpment. In fact, we strongly suspect that all of the cliff faces, especially along the western side of the Hudson Valley, are the products of glacial plucking. All those wonderful Shawangunk cliffs that the rock climbers and hang gliders enjoy are, it would seem, gifts of the Ice Age.

Figure 3-5. Fractured rock at Minnewaska State Park.

The old carriageway continues to the southwest and joins another red-marked path, the Gertrude's Nose Trail. That's the harder route, but it offers more and better geology. Follow it southwestward and you will see much more evidence of polishing, scouring, and plucking. This vicinity took quite a beating during the Ice Age. All this is manifest at Gertrude's Nose itself, which is the great cliff at the end of the trail (fig. 3-6).

We don't know who Gertrude was, but she must have had quite a nose. This extraordinary point of white sandstone was shaped during the Ice Age. It's badly broken up by more fractures, and at the base of the cliff there are many very large blocks of rock that have tumbled down the cliff face since the end of the Ice Age. Glaciers hadn't done all the work here; plain old gravity helped quite a bit.

Along the way are some more glacial features—a host of **glacial erratics** (fig. 3-7). Erratics are boulders that have been transported by advancing glaciers to the sites where they are found today. All along the ledge just west of the nose are a number of big and small erratics. One is truly large; it's called Patterson's Pellet (fig. 3-8), and it is perched on

Figure 3-6. Gertrude's Nose at Minnewaska State Park.

the edge of the cliff above Palmaghatt Kill. That's a curious thing that happens from time to time. Large boulders were transported south to the very edge of a cliff and then, with the final melting of the ice, they were left there just before they would have tumbled over the edge.

The Gertrude's Nose Trail turns and joins the Millbrook Mountain Trail, and that trail heads back toward Lake Minnewaska. Except for one steep climb, it's an easy walk. You continue on the red trail until you reach the lake. If you follow this route, you will have had a very nice jaunt and learned a lot of glacial geology along the way. And that will be the strategy we will use throughout much of this book. We are going to take you for a lot of similar hikes. It's a very nice way to learn geology.

Figure 3-7. Erratics along the Blue Trail at Minnewaska State Park, near Gertrude's Nose.

Figure 3-8. Patterson's Pellet, on the Blue Trail at Minnewaska State Park.

The Gunks are among the best locations to examine evidence for the advance of the ice, and what you learn there can be applied to the whole Hudson Valley. Wherever we travel, here and there, we find patches of exposed bedrock that occasionally reveal striations, crescent marks, and erratics. These are common throughout the whole region. And we mean up and down *all* of the Hudson Valley. Even New York City's Central Park is famous for its many outcroppings that display all these features. Collectively these abundant features comprise the heritage of glacial advance. But the irony is always that the ice destroys much of the evidence of its own advance; we have to search out what is left.

Fortunately, there is more evidence that helps flesh out the details to the story of the advance of the ice. All along the **Wall of Manitou**, the **Catskill Front**, we find cloves or breaks in the great wall where glacial ice overtopped the wall and entered into the Catskills. North/South Lake State Park shows a fine example right at the crest of the escarpment, just east of North Lake. The bedrock along the eastern shore of the lake displays striations with compass directions pointing west. Apparently, as the ice advanced down the Hudson Valley, it thickened, filled the valley and then swelled westward into the interior of the Catskills. The Laurentide advance was getting bigger. That's a topic that needs more discussion.

4

The Wall of Manitou

There is a fundamentally simple question that we ought to ask early in this book: was there a Hudson Valley glacier? We found out in chapter two that there was a great mass of ice called the Laurentide ice sheet that covered most of North America. It advanced out of the north and covered the entire Northeast. But was there also, perhaps a little ahead of it, a smaller glacier that we can call the Hudson Valley glacier? Did this smaller glacier lead the way and fill the Hudson Valley before the arrival of the great Laurentide ice sheet? We argued in the preceding chapter that there was a Hudson Valley glacier. Now we can try to pursue that question along the Wall of Manitou.

The Wall of Manitou—the wall of the Algonquian Great Spirit or God—is also called the Catskill Front. Topographically it is a ten-mile-long, straight-as-an-arrow escarpment (fig. 4-1). It rises a little more than 2,000 feet above the floor of the Hudson Valley. It has impressed and puzzled generations of geologists who have stood on its crest and wondered how exactly it came to be. We think that our hypothesized Hudson Valley glacier helps to answer that question and, in so doing, proves itself to have actually existed.

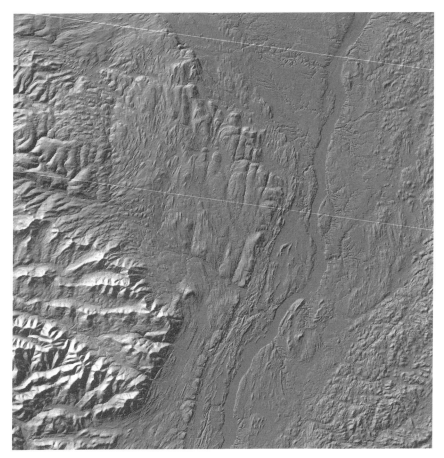

Figure 4-1. Wall of Manitou, satellite image. Courtesy U.S. Geological Survey.

We need to take a look at the geological evidence, and that starts with cracked rocks. If you spend time along the Catskill Front, you will soon observe long, smooth, straight fractures in the bedrock (fig. 4-2). Being essentially vertical, they have produced many striking cliffs all along the Wall of Manitou. These fractures are called **joints**. They formed perhaps 200 million years ago as part of the damage done during late-Appalachian tectonics.

At one time, Africa collided with our eastern seaboard. You can imagine how much compression that involved. When Africa eventually

Figure 4-2. Joints on Catskill Front. Widest one, foreground, just left of center.

broke free and drifted back off to the east, those compressed rocks experienced a release, and that is when they cracked to make the joints—lots of them; hike around and see how many you will find.

There are several sets of joints, but the most prominent set has a compass direction of about south, 40 degrees west. As it happens, that is just about exactly the compass direction of the Wall of Manitou. We think we can fairly deduce that the long straight joints have a lot to do with the long straight Catskill Front. We need an agency to link the two, however. Something had to turn the joints into the wall, and that agency was, we think, the passing Hudson Valley glacier.

Imagine a flow of ice moving southward through the Hudson Valley. Then imagine that flow thickening and rising up the Wall of Manitou, acting much like rising floodwaters in a river. The south-moving ice would roughly abrade that wall. But something more important than simple abrasion would have been going on. That something else would have been the plucking process we described in the last chapter. Ice forms a bond with the rock it passes by or passes across. The ice sticks to the rock, moves south and tugs on the rock; the rock resists, and "something's gotta give."

It's the joints that provided the "give." Eventually the rock gave way, fracturing loose along its own joint planes. Large masses came to be yanked free and carried off. Later, more rock was plucked loose, and over time we can imagine this process shaping a high, tall, very straight escarpment on a bearing of south, 40 degrees west. Smooth, vertical, joint surfaces would be expected to be common on the lip of the Catskill Front, and indeed they are.

But there is still more. If there really was a Hudson Valley glacier, then it can be expected that it thickened, expanded, and rose up the Wall of Manitou until it reached gaps at or near the top. At that point the ice should have flowed through those gaps and, later, across the crest of the Wall. Then, these fingers of ice should have flooded into the Catskills. At some point the great Laurentide ice sheet should have caught up with the local valley ice, and all the flow of ice should have been diverted to the south. And that's exactly what we see evidence of in any number of locations.

Our favorite such example is at North Lake. There, along the eastern lakeshore, you can find westward-directed striations (fig. 4-3) that record the flow of ice out of the valley, across North Lake and into the Catskills. That flow, incidentally, was very sizable; it scoured out the basins of both North and South Lakes. Watch for numerous striations along the rocky east shore of North Lake.

Similarly, along the green trail that rises from the top of Plattekill Clove and takes you up onto Plattekill Mountain, you can find another set of westward-directed striations. Evidently ice flowed out of the Hudson Valley and up into the higher reaches of Plattekill Clove.

The Green Trail provides some very interesting information. As you ascend the trail, which has a lot of striated bedrock outcroppings

Figure 4-3. Striations along North Lake shore.

exposed along the way, you will find that the striations start out bearing northwest, parallel to and within the confines of the clove. But later, at higher elevations, the striations shift to the west. If you climb to the top of Overlook Mountain, you will observe striations that are oriented to the southwest. A history of glaciation is recorded: a branch of the Hudson Valley glacier flowed northwestward into Plattekill Clove—at first. But then, as this flow rose, it stopped being funneled by the clove and veered westward. Then it was swept to the south as the Laurentide ice sheet caught up and took control of all the ice. The transition to a nearly western flow is at about 2,300 feet in elevation. The top of Overlook, with its southwest striations, is a bit more than 3,000 feet.

The great straight Wall of Manitou and the striations of its cloves speak, with one voice, of a single simple story. There was a Hudson Valley glacier, and it did sweep southward down the valley, steered mostly by the western wall of the valley. As it thickened, it began to flow up and over the Catskill Front and onward toward the west. But the Hudson Valley glacier was just the advance guard, a harbinger of things to come. With time, a greater flow of ice caught up and replaced it. That was the main act, the Laurentide ice sheet. Its ice spread across both the Hudson Valley and all of the Catskills. It also would have covered all of the Taconics and Berkshires to the east.

That was a lot of ice! It was heading south. How far would it go?

5

The Great Moraine

Let's return to an earlier image. Imagine for a moment that you lived in a world where great ice sheets covered much of the globe and they were still advancing. Let's make it worse—let's suppose that you lived just south of an advancing ice sheet. Then you would have one overriding question: how far south will the ice advance? Or, more to the point, will the ice get to *me*?

If you lived in the Hudson Valley back during the advance of the ice, those would be very pertinent questions. We don't live in such times, of course—the glaciers stopped advancing long ago—so the answer is history, and it is available to us now. The ice reached what is today the northern half of Long Island. Take a look at the map (fig. 5-1); it presents a fascinating revelation about the very nature of that now densely populated island. It turns out that Long Island is another gift of the Ice Age.

The Laurentide ice sheet kept advancing until the rate of melting exactly balanced the rate of advance. In other words, snow from Labrador was forming ice that pushed south and advanced as far as northern Long Island, and then and there it melted. The formation of ice in the

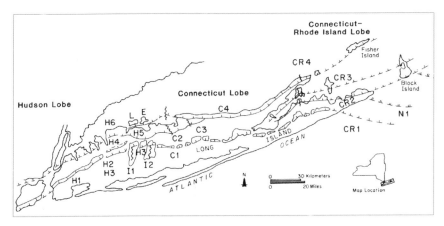

Figure 5-1. Map of Laurentide Ice Sheet at Long Island and location key. Black bands are the terminal and recessional moraines. Courtesy New York State Museum.

H1 Harbor Hill Moraine
H2 Jericho Moraine
H3 Old Westbury Lobe
H4 Oyster Bay Moraine
H5 Northport Moraine
H6 Sands Point Moraine

Interlobate Moraine

I1 Manetto Hills Lobe
I2 Dix Hills Lobe
I3 South Huntington Lobe
I4 High Hill Interlobate Moraine

Connecticut Lobe Ice Margins

C1 Ronkonkoma Moraine–Shinnecock Moraine
C2 Stony Brook Moraine
C3 Mount Sinai Moraine
C4 Roanoke Point Moraine

Connecticut Lobe and Eastern Connecticut–Western Rhode Island Ice Margins

CR1 Amagansett Moraine
CR2 Sebonack Neck—Noyack–Prospect Hill Morainal Envelope
CR3 Robins Island—Shelter Island—Gardiners Island—Morainal Envelope
CR4 Roanoke Point—Orient Point—Fishers Island Moraine

Narrangansett Lobe

N1 Montauk Point (Altonian)

Labrador ice center was exactly balanced by the melting to the south at Long Island. There was an equilibrium that maintained an ice front along the northern edge of the island.

This dynamic created a conveyer-belt effect. As it moved to the south, the ice scraped up a great deal of sediment—clay, silt, sand, gravel, cobbles, and boulders. The sediment was swept along and carried south, concentrated at the base of the advancing ice. When the ice and its sediment load reached that Long Island terminus, the ice melted and the sediment was left behind, dumped unceremoniously in a heap. More advancing glacial ice replaced what had been melted, and it brought along still more sediment. Over time the conveyer belt carried more and more ice and sediment to the glacier's front. The ice always melted, and meltwater always flowed away, off to the south. But the sediment piled up in an increasingly large heap called a **terminal moraine**.

The terminal moraine at Long Island runs east to west across the whole island. It also extends westward into northern New Jersey and northeastern Pennsylvania (fig. 8-1). In fact, it can be traced across all of the United States (bottom of shaded area on fig. 2-2). It is the Laurentide terminal moraine. On Long Island it is called specifically the **Ronkonkoma moraine**.

Wherever roads have cut through a great moraine of this sort, you will see large quantities of coarse-grained sediment exposed (fig. 5-2).

Figure 5-2. Moraine sediments exposed along Rte. 32, south of Cairo.

The presence of *very* large boulders is a diagnostic feature of such moraines. The landscape of a moraine is very striking; typically it is a sinuous rolling "sea" of small hills and dales (fig. 8-3). These are called **kames** (hills) and **kettles** (dales). It's a handsome landscape, especially in the late afternoon when sunlight and shadows highlight its features. But it is nearly invisible when forested, which is in most cases.

The kettles mark the locations of what had been large masses of ice that had come to be buried within the moraine. These were probably fragments of the ice sheet itself that came to be detached in the melting process and then buried. Being under all that sediment, these land-locked "icebergs" were well insulated. They may have lain buried for decades, or possibly even centuries, before finally melting. As they thawed out, the overlying sediment collapsed to form the kettles. If the kettles were deep enough, they were filled with water and are called **kettle ponds** or, if larger, **kettle lakes**. These can be common, and small ponds in our region are frequently kettle ponds (fig. 5-3).

South of the moraines, the ancient ice age dynamics are revealed in a very different sort of landscape. At the time, when there was melting ice

Figure 5-3. Kettle pond near Red Hook.

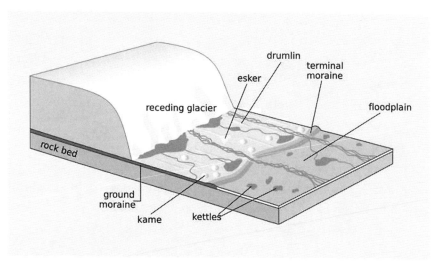

Figure 5-4. Moraine/outwash complex. Advancing ice brings sediment to the front of the glacier. After melting, a moraine and an outwash plain are left behind. From Luis María Benitez in Wikipedia.

---

just behind the moraine, there were enormous volumes of water flowing across the moraine's sediments. Much of that sediment, especially the fraction that was sand, came to be washed south. It accumulated as a sheet of mostly sand that is called an **outwash plain**. We call the whole deposit a **moraine/outwash complex** (fig. 5-4).

That outwash plain is exactly what it sounds like—it is mostly a smooth and flat plain composed of sand that had washed out of the moraine. Much of the southern half of Long Island is a very broad outwash deposit; after all, a big moraine will produce a large outwash.

We said that the outwash is smooth and flat, but there are some important exceptions. There are kettles, and commonly kettle ponds, on outwash plains. These formed just as they did within moraines as described above.

Take another look at the map of Long Island. To the north of the terminal moraine, especially in the central and eastern part of the island, is a second moraine. It speaks to us of "the beginning of the end." The terminal moraine was an active landscape feature as long as large amounts of ice were advancing out of the north. But, if the climate changed and less ice was provided from the north (Labrador), then the advantage would swing over to melting. More ice would melt than could be replaced by glacial advance. Now the front of the glacier must have been forced into a retreat. The ice itself could not actually *move* backward, of course, so it is better to say that the front of the ice *melted* back.

Again, imagine if you lived during the middle of an ice age. You have witnessed the advance of the ice for what has seemed to be eons and now, at last, it is retreating. What joy such a moment would bring! But such glee would likely be temporary; it seems there is no such thing as a smooth, continuous retreat of the ice. Instead the process of deglaciation might be best described with the phrase "two steps backward and one step forward." That second moraine of northern Long Island is the one-step-forward moraine. It can be called a **recessional moraine**. It records a chapter in the Ice Age when, once again, the climate had become conducive to the advance of the ice. What a frightening moment that would be for anyone who happened to be there. Enough with the ice already!

Take another look at that map. If you are ever traveling across Long Island, keep it in mind. Wherever there is some relief to the landscape, especially those hills and dales, then you are likely driving across or through a moraine. If, however, you are on a broad flat landscape, then you are probably driving across the outwash. Remember: moraines to the north, outwash to the south—that will help a lot. Keep a sharp eye out for holes in the ground, especially if they are filled with water; they are likely to be kettles. The largest is Lake Ronkonkoma, at about three-quarters of a mile across. There are about seven other large kettles and scores of small ones on Long Island. The area around Middle Island is especially known for its many kettle lakes. Artist Lake, Spring Lake, Bartlett Pond, and Upper and Lower Lakes are all excellent examples of kettle lakes in that area. All these are features that we will see again, for they are found up and down the Hudson Valley as well. And we will visit full-fledged moraine/outwash complexes again, too—a number of them.

6

The Hudson Canyon

It would be a strange thing for us, the mind's eye, to watch the progress of the Hudson Valley glacier during the many centuries of its long advance to the south. We would see the ice always stretching out as if to reach the sea at the end of the Hudson River, but we would see our glacier always falling short of reaching its oceanic goal. The glacier was like Sisyphus, who always attempted to roll his boulder to the top of a hill only to watch as, at the last second before reaching top, it rolled back down to the bottom. Our Hudson Valley glacier would be similarly frustrated, as the Atlantic Ocean would always seem to shrink away from its advancing front. The farther south the ice advanced, the farther south the waters of the ocean would retreat. The glacier, it would seem, would never reach the sea. Why?

The answer is simple, but important. As a great ice sheet expanded in size, it consumed water; it trapped any snow or rain that fell, converting it into more ice. So much water was becoming ice that the only places where water could be found in sufficient volumes were the oceans. As the glaciers grew and advanced, the level of the oceans all around the world dropped and the shorelines shrank away, in this case to the south.

Oddly, as the northern and middle stretches of the Hudson Valley disappeared under the advancing ice, new expanses of the valley opened up to the south, exposed by retreating seas. Whatever Hudson River there was, at this frozen time, came to flow out onto what had hitherto been the continental shelf, the shallow bottom of the Atlantic Ocean just offshore of today's New York City.

Similarly, the whole coastline of the Atlantic Ocean retreated in the face of the advancing ice, and for the same reason. Sea levels were dropping here and around the whole world. That is a counterintuitive notion to a person living in the world as it is today. We live in an era when glaciers are melting and sea levels are slowly rising. Thus we think that rising sea levels are the norm, but that is true only for our time; it was not true 21,000 years ago.

There was a moment in time when the Laurentide ice sheet had reached its all-time maximum and, conversely, when sea levels had dropped to their all-time lows. We can't place an exact date on that time, but it was probably about 21,000 years ago, give or take a few thousand. We do know that, back then, sea level was about 450 feet lower than it is today. We have been there already; when we imagined ourselves as the mind's eye in the introduction to this book, we flew out across what is today the bottom of the Atlantic Ocean. Down below we saw a dry landscape, and a lot of it. We know that surface today as the ocean's continental shelf, but back then it was a broad, flat, sandy coastal plain. It has been reported that, on rare occasions, fishermen out on the Atlantic have pulled mastodon teeth up in their fishnets. Those teeth are silent testimonials to a very different time and a very different ecology.

But what about the Hudson River? As the ice was expanding, the climate must have been quite frigid. That river's channel was likely empty much of the time. When it was thawed out and flowing, it must have labored to carve a channel across what was then a new coastal plain. That effort actually may not have been all that hard; the seafloor had been composed of soft sand, and it is relatively easy to erode a channel into that sort of material. This channel has a name; it's called the Hudson Shelf Valley.

Oceanographers have, in fact, located this old channel, and it does cross what is now the bottom of the sea (fig. 6-1). Near the modern mouth of the Hudson River, the channel is about a hundred feet deep. The Hudson flowed across dry land when it carved that stretch. But when those oceanographers followed that canyon about 100 miles in an offshore direction, they found that it developed into something that was far more interesting and, at first, very enigmatic. It broadened and deepened to become a canyon large enough to rival the Grand Canyon—a seafloor feature called the **Hudson Canyon**. Today it is a very deep submarine canyon that cuts three-quarters of a mile into the continental shelf and also into the continental rise beyond. But, when it was discovered, this posed a really puzzling problem—how could such a deep canyon be located at the floor of the ocean? "Grand Canyons" should be on dry land. That was a real quandary.

The evidence that led to an understanding of what happened appeared in November 1929. An earthquake struck in the vicinity of Newfoundland. Over a period of thirteen hours, a series of trans-Atlantic cables snapped one by one in a sequence stretching from near the earthquake's epicenter to hundreds of miles offshore in much deeper waters. Evidently, the earthquake had generated a submarine avalanche. This was an avalanche not of snow, but of sediment and water, and it thundered down the marine slope at as much as sixty miles per hour; it was this that easily broke the cables. Geologists pondered this event and then applied what they had learned to the Hudson Canyon. They cracked the problem.

We have to imagine what it was like for a substantial river to cross an emergent continental shelf, probably at about the time when the glaciers were just starting to melt. A swollen river composed of huge amounts of meltwater was carrying vast quantities of sediment, and all of it was ending up at the edge of the emergent and dry continental shelf. The river would disgorge all this sediment at its mouth, which was at the top of the continental slope—the steep-sloping part of the seafloor beyond the shelf. As the sediment piled up there, it became over-steepened. That means that the excess sediment made the seafloor slope too steep to be stable. On land this situation would result in frequent landslides.

We do, in fact, get a similar phenomenon on the steep floor of the ocean; it is called a **turbidity current**. You might think of it as a submarine avalanche. Periodically, something, very plausibly an earthquake, jars the seafloor and triggers the avalanche. As its sediment-laden waters rush down the slopes and into the depths of the sea, they erode an ever-

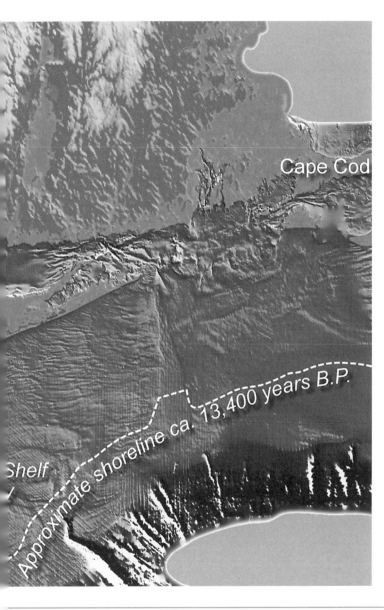

Figure 6-1. The Hudson Shelf Valley and Hudson Canyon. Illustration by Jack Cook of the Woods Hole Oceanographic Institute.

deepening and widening canyon. That's what happened offshore of Newfoundland, and that is what is thought to have happened at the bottom of the Hudson Canyon. Over sufficient lengths of time, such canyons can become enormous. We say "canyons," plural, because similar ones are found offshore coastlines all around the world. Our Hudson Canyon is a very good one; it is the largest one, of many, along the east coast. And it has one distinction that differentiates it from all the others. It may be, in part, the product of a separate event. A great flood is also thought to have contributed to the formation of this enormous canyon at the very end of the Ice Age. This is the flood associated with the sudden drainage of Glacial Lake Iroquois. We will return to this topic and elaborate on it in chapter thirty-four.

At the bottom of the Hudson Canyon, there is quite a heap of sediment, and it all had been transported down the ice age Hudson River. Material that had once been part of the Hudson Valley is now under thousands of feet of water.

The discovery of these post–ice age submarine canyons came as quite a surprise to early-twentieth-century geologists. These canyons should certainly be regarded as one of the most important categories of ice age features ever recognized but, sadly, few people can go and look at them. They are so deep and so far offshore. But, as we return upstream, we will have a lot of good glacial features that we *will* be able to see.

The First Retreat

We must return, once again, to that critical moment in time that marked the beginning of the end. There must have been such a moment, a wintry hour on a wintry date when the Laurentide ice sheet had reached its absolute zenith. That was the exact moment when the volume of the ice was at its greatest. It was the day when the elevation of the ice was at its highest. It was an hour when sea level had ebbed to its ice age bottom. Somewhere there was the lowest of low tides, and mudflats that had never been exposed before, nor ever would again, felt the dim heat of the ice age sun. This moment in time marked the Ice Age at its absolute peak.

The mind's eye, orbiting above our planet, would have seen that it was winter in the northern hemisphere, and we would have seen a low sun "shining" down toward the North Pole of the rotating globe. Most of the sunlight missed its mark and left that pole mostly in darkness. Some of the sunlight, however, did reach out across the northern Pacific, and it would have reflected off much of the vast expanse of the sea ice that spread out across the whole top of planet. That included a great deal of floating ice that extended out from Siberia's Pacific and Arctic coasts. In the dim winter light, Siberia itself was mostly a frigid and frozen desert with a huge expanse of tundra. As we watched in fascination, the planet slowly rotated below us, and more of Siberia came into view. In its western regions the great cold landscape displayed a huge ice sheet that, as it rotated into our line of sight, expanded in size. It reached all the way into the darkness of the North Pole and as far beyond that pole as could be faintly seen in the starlit darkness.

The Earth continued inexorably in its rotation below us, and so it was that the great expanse of the ice continued to be revealed by the cold light of the Earth's perpetual dawn. All of Scandinavia was glaciated, as too was all of the Atlantic from Iceland to the north. Now Greenland and then North America rotated into the morning sun to take their turns in this Arctic carousel. Greenland was entirely covered with glaciers, but much of North America had managed to escape the ice. Still, the whiteness had expanded south all the way to what are the coastal regions of today's Long Island. Just to the west, other lobes of ice had expanded into northern New Jersey. The white shroud continued into the Midwest, where its advance had encountered easy going on the low, flat landscape of the upper Mississippi Valley. Beyond, the glaciers had found the mountains of the American West to be a difficult struggle. Their southward advance had there been impeded by the rugged peaks.

There is something fundamentally unstable about an ice age on the planet Earth. Such a time is, after all, a climatic extreme. So it was that this zenith of the Ice Age was just a moment in time. Somewhere the temperature, in that morning's sun, went up by just one degree, but that marked the beginning of the end. Slowly the climate would begin to warm and the ice would retreat from its fleeting triumph.

Our global merry-go-round ride carries us around the Earth one more time and brings us back to the focus of our story, and that is a narrow one, confined only to the Hudson Valley region. With the beginning of the end of the Ice Age came a start of the retreat of the valley's ice. With retreat, the glaciers would leave behind landscape features, prominent among which were the moraines. These would form earthen dams, and those would impound vast quantities of meltwater. This was

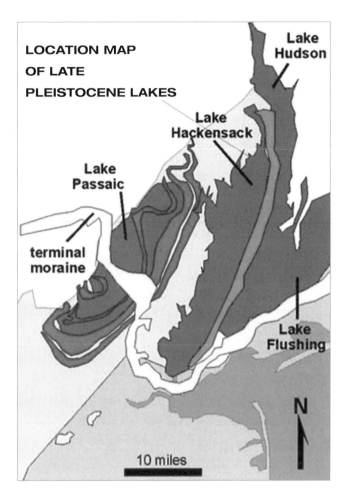

Figure 7-1. Map of the ice age lakes in the lower Hudson Valley and New Jersey. Courtesy U.S. Geological Survey.

the forming of the first of many glacial lakes. These earthen dams would create a New York metropolitan region altogether alien from the one we know today. It was a landscape of sizable lakes—ice age lakes.

It cannot be much of a surprise to find that, at a time of melting ice sheets, there was a lot of water available to create lakes. In the early going there were several large ones at the southern end of the Hudson Valley (fig. 7-1). These were Glacial Lake Flushing, east of New York City, and Glacial Lake Hackensack, west of the Hudson. They were, primarily, occupying low-lying lands just north of the terminal moraine. Glacial Lake Passaic was surrounded by today's Watchung Mountains in northern

New Jersey. Glacial Lake Connecticut lay to the east, behind the Harbor Hill moraine. These were big lakes, thanks to the Ice Age. They are now mostly meadowlands, soggy islands of low population occupying large swaths of land in the heart of one of the world's most populated regions. These were only the first of many lakes that would be created by the Hudson River glacier. We will travel up the valley and see a good number of them.

8

Later Retreats

The retreat of a glacier is a pretty tricky business. We have already described the basics (see chapter five). Again, the first thing that you have to appreciate is that the ice did not ever *literally* retreat. It did not actually move to the north; it couldn't—nothing could have pushed it in that direction. In fact, the bulk of the ice was almost always moving south, even during a retreat. You see, during the waning times of the Ice Age, up in Quebec and Labrador, the snow was still always piling up, turning into ice that expanded out from its center of accumulation. There was always a shove to the south, a stress that was transported through all of the ice, forcing all of it forward. And this was likely the case right up to its southernmost reach, where the ice melted all along its periphery.

So, despite a pattern of retreat, the ice was usually still moving forward. As long as the melting fell short of the rate of advance, the ice front edged onward; however, when melting exceeded advance, the front of the glacier retreated. What we call a retreat was actually caused by the melting, which would have been seen mostly along the front of the glacier.

Imagine standing at the front of a disintegrating glacier. If you are patient, you can wait long enough to occasionally hear cracking and groaning sounds emanating out of the glacier as it lurches forward. But mostly you would be watching enormous amounts of meltwater pouring off of and out of it. Occasionally you would be treated to the image of a great mass of ice collapsing in front of you. The tricky thing to remember is that the ice is still almost always moving forward.

Confused? It gets worse; there were often times when the melting and the advance of the ice were balanced. The glacier's front then became stabilized along a specific line. The ice constantly advanced to that line, and there it melted. More ice reached that line, and it also melted, and then it all happened again—you get the picture. Such stable fronts of a glacier are called **stillstands**.

When we saw this on Long Island, we were looking at the Laurentide ice sheet. That huge glacier had overrun both the Hudson Valley and the adjacent Catskills, along with the nearby Taconic and Berkshire Mountains (and most of North America). This formed a glacial front that extended across the whole continent. But, when we are looking at the retreat of the ice within the Hudson Valley, our focus will be far more confined. Pretty much what happened for thousands of years or so can be described as follows: first the ice would retreat for quite a distance as climatic warming and melting exceeded the rate of ice accumulation; then things would cool down for a while and the front of the ice would stabilize during that interval. We describe that line of stability as having been an **ice margin**.

See the map in fig. 8-1. All the black lines are ice fronts, and each represents an interval of stability. Each represents a relatively short interval when the ice front was stable, and each interval of stability was followed by a retreat. These are what we call ice margins. All the ice margins are named for towns where their features are well preserved. The patterns reveal that the glacier had always retreated more slowly in the *center* of the Hudson Valley than on the mountaintops where the ice was thinner. Also, toward the eastern and western edges of the valley the flow of the glacier was retarded as the ice dragged against higher landscapes. The ice margin lines reflect this. Notice, too, the very strongly concentric

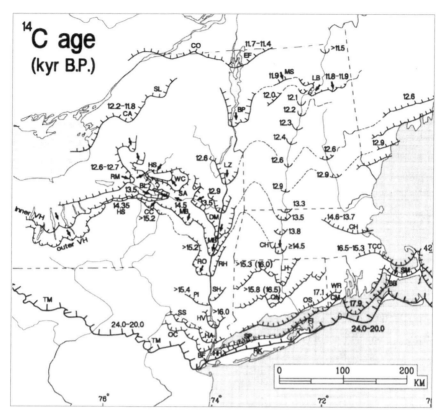

Figure 8-1. Ice margins in the Hudson Valley. Each records a pause in the retreat of the ice. Courtesy of the New York State Museum.

nature of these ice margins. Each later retreat would be parallel to the most recent earlier ice margin. The result, up and down the Hudson Valley, is this strikingly concentric pattern.

But there is more. During the complex retreat of the Hudson Valley ice, there was also a series of events that might be described as "two steps backward and one step forward." It might even have been that the climate cooled down so much that, instead of just stabilizing, the ice actually expanded southward again. We call such events **readvances**. Such a readvance will always be followed, eventually, by renewed retreat. Those were the two steps backward. Some of the heavy black lines on the map are the readvances.

We can't help but look at a map of this sort and think about what we cannot see. These well-defined concentric lines reveal so much about the retreat of the ice from the Hudson Valley, but at the same time they conceal a record of the advance of the same glacier.

Remember that advancing glaciers destroy most of the evidence of their own forward motions. We can't help but imagine that a similar set of lines might be drawn to document the advance of the Hudson Valley glacier, but there is no record of that advance because the glacier destroyed almost all records of such movement. The ice advanced and, at the same time, wiped out all evidence of that advance.

We only see a record of retreat, but the interruptions of that retreat left a sedimentological record, and a good one. Ice margins, whether stable stillstands or readvances, produce predictable sets of features. What are those features? They are the same sort of moraine/outwash complexes that we saw on Long Island, but they are generally smaller in the confines of the Hudson Valley. The moraine/outwash complex on Long Island is enormous; none in the Hudson Valley is anywhere near as large. Large or small, moraine/outwash complexes are very recognizable, even striking, features. The moraines display their kame and kettle topographies, and these grade, immediately, into the flat outwash plains, which also may display kettles.

The best example of a moraine/outwash complex that we have found, so far, in the Hudson Valley lies just east of the village of Red Hook. Take State Highway 199 about 1.5 miles east of town, and in the Yantz Road vicinity gaze north. The rolling hills on the horizon are very good kame and kettle landscape (fig. 8-2). This is the Red Hook moraine. Below this moraine are the flats of a very fine outwash immediately north of Rte. 199. We passed Yantz Road and turned north on Crestwood Road. This took us right into the heart of the Red Hook moraine, and the road cut right through it, giving us a very good look. The kame and kettle landscape was especially well developed (fig. 8-3).

We have looked, but we haven't yet found any moraine complex elsewhere in the valley as well developed as this. We have found some others, but not quite as good. We turned around, returned to Red Hook and turned north onto Rte. 9. For about five miles along the highway, we were

Figure 8-2.
The Red Hook Moraine. Moraine in distance, outwash in foreground.

Figure 8-3.
The Red Hook Moraine, kames and kettles.

able to trace a pretty good moraine, mostly west of the highway.

On another day we followed Columbia County Rte. 7 north from its intersection with Columbia County Rte. 2. We found what appeared to be another moraine just past the intersection. That moraine was much better exposed as we passed by Hall Hill Road and approached Ancram. This moraine continued to be well exposed most of the way to West Copake. There appear to be three moraines in this part of Columbia County, and all three are oriented at a roughly southwest-to-northeast parallel. Our favorite, however, is the one east of Red Hook. We came back for another look.

September 18th, 18,475 BP. It has been a very warm summer, and the Hudson Valley glacier has been stagnating along a front that passes through what will someday be the village of Red Hook. The glacial advance that had brought the ice here came to a halt just a few years ago. We look north and see the glistening image of heaps of very wet sediment. These rolling hillocks are devoid of any foliage; few plants have been able to establish themselves in this still mostly ice age ecology. Because of that, we can see the sediment quite clearly. There are a very large number of enormous boulders within it. The glacier, when it was advancing, was quite capable of picking these up and moving them along. Now they have come to rest. Two of these boulders are composed of a complex crystalline rock called gneiss. They likely came from the Adirondack, the Taconic, or Green Mountains, which are the closest sources for such mountain core lithologies (rock types). These rocks are not native to Red Hook. The rest of the moraine sediment is a complex of cobbles, gravel, sand and stuff (fig. 5-2). Here and there, large masses of dirty ice can be seen poking through the sediment.

Rising above and behind the moraine, and also above the entire northern horizon, is the glacier. It reaches to a level of a hundred feet or so above the moraine. It, too, is very wet and it, too, glistens. It is melting today as it has been all summer. On this gray cloudy day it is a dark beryl-blue color, and it is broken up by many large, dark fissures. The ice is melting along these fractures, and they have widened considerably.

Large amounts of water have been flowing from many of these fissures, and that water is organizing itself into shallow and broad streams that flow down the slopes of the moraine. We watch as some of these flows pass right by us and then, following the flow of the water, we turn around and look south. There before us is another glistening landscape, but this one is flat and low. The streams flowing across it are numerous, and they form a crisscrossing pattern that is known to geologists as a braided stream complex (fig. 1-1). This is typical of an outwash plain, and this is how sediment is brought to an outwash and spread out across its flat surface. And, in fact, this is why that surface is flat (fig. 8-2).

We slowly turn a full 360 degrees and once again absorb this panoramic image of a waning ice age and a retreating Hudson Valley glacier.

9

The Drowned Lands

The era of glacial retreat produced some fairly odd landscapes in the Hudson Valley, and none may be stranger than those found in the Wallkill Valley of Orange County. There is a stretch of landscape right on the New Jersey border that is called the "**drowned lands**." We think, as we will explain in chapter twenty, that it should be called the *southern* drowned lands. Sometimes this landscape is called the mucklands, or the black dirt region, and both of those names might well represent better choices of words, more precise in their descriptiveness. Nevertheless, these lands are most definitely "drowned."

Today it is a landscape famous for its agriculture; it is an excellent place to grow onions. Celery is also harvested here, as is sod. In the spring, when plowing has been finished, the soils are well exposed and seen as black mucks because they are so wet. These drowned lands are no accident; they take us back to an interesting and unusual moment in the deglaciation of the Hudson Valley. They are important consequences of ice age history.

As we begin our description of the drowned lands, we want to make an important point—perhaps even to establish an important theme for this book. We want to emphasize that we envision the many landscapes of the Hudson Valley at the time immediately after the ice melted as being really, really wet. When we travel throughout the Hudson Valley, we look into the past and we see, in our minds' eyes, numerous water-logged vistas. We see glistening, soggy landscapes. Pools of water appear all over. There were numerous, mostly shallow, lakes and ponds, and swamps, bogs, and marshes. You name it—if it was wet, it was found here. Such streams as did exist in the valley bottoms were often sluggish. Drainage could be enormously inefficient. There were, at best, only limited patches of non-aquatic vegetation on the higher elevations of these landscapes, still known today as islands. The soils were just too soupy for plants to thrive; their time would have to wait. Elsewhere in the Hudson Valley and Catskills, on sloping landscapes, there were pounding, thundering, powerful white-water streams, but we are speaking now of the drowned lands; they were still, silent, and just plain wet. And they were very common throughout the Hudson Valley.

The drowned lands of Orange County represent what were the wettest of such terranes. This was the largest glacial lake outside of Lake Albany (chapter ten). It was sizable, but shallow. Its origins date back to the retreat of the ice. At about 18,400 years ago the retreating ice created the Pellets Island Moraine. This heap of earth stretched across the Wallkill River Valley, whose waters would have otherwise flowed to the northeast, as they do today. The waters of the Wallkill were, at least in part, impounded by the moraine and the ice lying behind it. To the northeast were the Shawangunk Mountains, and they blocked any flow of water in that direction. To the southeast was Mt. Peter and it, too, acted as a dam. All this dammed the Wallkill River and created the lake, called (logically enough) **Lake Wallkill**, although it might be better to describe it as a giant pool of water. Such a still pool of water would likely have been stagnant, and that is the beginning of its history.

Lake Wallkill must have been a sizable body of water right after it first formed. The muck and peat deposits that fill its basin today are as much as thirty feet deep. But size is not the most important aspect; the stagnant nature of the old pool is what counted. Over long periods of time, a pool of

Figure 9-1. The black peats of the drowned lands.

this sort is said to undergo eutrophication. Plants would have grown and died, especially around the edges and near shore areas of the water. With time vegetative material began to accumulate, some of it alive, much of it dead. A mat of floating vegetation began to expand out from the shore. This was the beginning of the formation of the black muck. In the stagnant waters beneath the floating mat were found ideal conditions for the accumulation of the organic-rich, fine-grained stuff that we call muck. It is black because of the very large amount of organic carbon in it. Over the millennia these mucks gradually filled in the old lake and created the drowned lands, a shallower, but still very wet environment.

These mucks were ideal for the preservation of the remains of ice age mammals, and more have been found here than anywhere else in the region. The list is a long one and begins with mastodons, dozens of which have been found.

Eventually, settlers arrived. They cut several canals and drained most of the wetlands. In so doing they established a rich farmland (fig. 9-1). There is, however, still an ice age heritage here that can be found in the geographic names. Today there are limestone hills that rise above the farmlands (fig. 9-2), but in the early nineteenth century these were islands, places to which you generally had to take a boat if you wanted to visit.

Figure 9-2. Mt. Adam and Mt. Eve, "islands" that rise above the drowned lands.

December 17th, 18,389 BP—On our visit we stand near what would some day be the village of Denton, and we look to the south. We are standing upon the Pellets Island Moraine, and that marks the shore of old Lake Wallkill. We see spread out before us the expanse of that very large lake. We can hardly see its other, southern, side. Then we turn around to look north, and in that direction we see the front of what should have been a retreating glacier. On this day, however, it is actually advancing. The climate had recently been cold, and the ice had moved forward for one last advance. In front of the glacier we see enormous piles of coarse sediment. We can occasionally see masses of it lurching forward, shoved from behind. Some of the coarse sediment even tumbles down its steep front, forming small landslides. This advance is but a temporary event; the climate will soon warm, and the ice will resume its retreat. It will leave its moraine behind to form the dam that drowns the drowned lands.

Our journey to the drowned lands had taken us to two very different times—the present and the ice age past—and two very different landscapes. We stand and see them both at the very same time; we are privileged to be able to do this.

10

Lake Albany

We would like you to try a simple experiment. Press one hand down on a bed mattress. Notice that all around your hand the mattress becomes depressed. The farther away from your hand, the less the depression is. Now pull your hand in one direction while continuing to press down on the mattress. Notice that the depression continues to be greatest close to your hand, but away from it the mattress springs back to where it started. The depression "follows" your hand around on the mattress. This simple process tells you much of what you need to know about the greatest of all the Hudson Valley's ice age features—Glacial Lake Albany.

We have seen that the glacial ice began to melt in the lower Hudson Valley some time after about 21,000 years ago. With time, the glaciers melted and the front of the great mass of ice began its northward retreat from the valley. That thick glacier weighed a great deal and, like your hand on the mattress, it pressed down on the bedrock of the crust beneath it. At the peak of the Ice Age, the ice pressed down on the crust of the entire Hudson Valley. As we saw earlier, the retreating ice stabilized at a number of locations (fig. 8-1) and left recessional moraines. But, as the ice retreated from each of these stillstands, its weight was removed and the crust south of the retreating ice began to spring back. Geologists have a technical term for such crustal behavior; they call it **isostatic rebound**.

Isostatic rebound is not a quick process. Your mattress will spring back almost instantly when you remove the pressure of your hand, but the crust of the Earth responds much more sluggishly. The rebound is very gradual, taking perhaps even centuries to accomplish. During that long time the crust just south of the retreating ice would remain depressed and formed, of course, a basin.

The Catskill Mountains must also have been depressed by the weight of the ice, and so, too, must have been the Taconics and the Berkshires. But such landscapes, elevated to begin with, would never, and could never, form water basins. The Hudson Valley was different; it became a prefect bowl, and it did fill with water. And there was a very large amount of water. The whole mass of the melting and retreating glacier was turning into meltwater and flowing off to the south into that isostatic basin. It filled that depression, hence the origin of a glacial lake. What we are portraying is a very large glacier retreating up the Hudson Valley with an equally large lake forming in its place. That lake was Glacial Lake Albany (fig. 11-7).

In the centuries that succeeded the melting of the ice, there must have been a very great deal of meltwater flowing into Lake Albany, and there must have been very strong currents flowing through the lake and heading south toward an escape into the Atlantic. Late at night in geology bars, that escape and its exact route has been debated for decades. Many argue that the flow passed through the Arthur Kill, just west of Staten Island. That location makes sense because the terminal moraine of the Laurentide ice sheet (chapter five)—lay upon Staten Island and stretched to Brooklyn and beyond to the east, blocking and damming the course of the Hudson River.

If this is so, and it probably is, then we can imagine a great earthen dam and a long, depressed Hudson Valley lying behind it to the north, filled with water. In its lower reaches, at today's Tappan Zee and Hudson

Highlands, Lake Albany probably didn't look too different from today's river, though perhaps a good bit deeper. But farther north of that, the lake would have been quite a bit more impressive, much deeper and wider than today's river. It probably averaged about ten miles across. It must have been sixty to eighty feet deep, or more. And it stretched all the way north to beyond Glens Falls. Those water depths are calculated by subtracting the elevations of the lake's deltas from those of adjacent lake bottoms.

The currents, flowing mostly south through the lake, must have transported vast quantities of sediment—largely silt, clay, and fine sand. Much of that sediment came to be deposited on what would become the floor of the Hudson Valley. Glacial lakes deposit a special type of sedimentary feature; their bedding is said to be "varved." **Glacial varves** consist of alternating strata. Some layers are thin, very dark, and very fine grained. These alternate with thicker, coarser, sandier, and lighter-colored beds (fig. 10-1). The dark horizons accumulated during the winters, when the lake was frozen over. During those times

Figure 10-1. Glacial lake varves, Saugerties.

the lake water became stagnant, and the finest sediments settled to the bottom of the lake. Black biological materials settled, too, and that is why winter varves are so dark. During summer months, when the ice melted, currents picked up, and that is when the coarse sandy stuff was deposited. These are lighter-colored and thicker varves. Varves can be very impressive deposits, and we have seen them when the individual couplets were as much as four inches thick.

Unfortunately, varved sediments are almost never exposed unless somebody has been excavating into them in the most recent of times. They soon are overgrown with foliage and quickly become invisible. So, it is hard for us to tell you where to look for them; all of the varved sequences that we have seen are no longer visible. We recommend a hike down the nature trail at the Esopus Bend Nature Preserve in Saugerties. That's the most recent place we saw some exposed varves. Watch the steep banks along parts of the trail.

The varved lake deposits are always horizontal; that's the nature of deposition in general. The sediments are spread out horizontally under the influence of gravity and lake currents. This accounts for a great deal of the flat landscape that you can see while traveling the length of the valley, especially along some of the area's major highways. What a person should look for in the Hudson Valley are long stretches of flat landscape. The bottom of the lake became blanketed with sediment, and the currents that brought the sediment tended to spread it out in wide, flat sheets called strata. These strata can pile up to great thickness, but always, at the top, these layers will be flat, and so too will the landscapes that they produce.

It's something that we want you to develop an eye for in the Hudson Valley. We recommend a journey along Rte. 23 and Rte. 9 in Columbia County. Head east about a mile and a half from the Rip Van Winkle Bridge and you will encounter a great deal of flat landscape. This, of course, is the bottom of Lake Albany. There is a short interruption where Rte. 9 branches off to the south, but soon more lake bottom appears. Several miles of lake are seen from Rte. 9, mostly to the west of the highway. There is another small patch of lake bottom at the village of Clermont.

There are similar landscapes on the western side of the Hudson. A journey along the New York State Thruway northward from the Kingston exit will pass a good deal of such flat landscape. Much of the landscape to either side of the highway is lake bottom until you pass the Saugerties exit. These two stretches of highway serve well as "introductory" lake bottoms, but as you get more experienced you will find much more lake all up and down the Hudson Valley. We will come back to the topic in chapter sixteen.

Lake Albany used to be of considerable economic value. Its clay deposits were often just the right consistency to be used for the manufacture of bricks. Sizable clay pits were opened up all along the Hudson, and factories were built along its banks to turn the clay into bricks, mostly during the nineteenth century. Some hobbyists search the banks of the Hudson for bricks labeled by their manufacturers, and some brands are prized more than others. The locations of many of the old clay pits are marked on state topographic maps, and we have scoured the landscape looking for some that still show exposures of the clays. Sadly, we have found none. Most have been landscaped and planted with grass to hide the damage done by the quarrying. Others have simply become overgrown of their own accord. We recognize that the harmful visual damages of quarry activity should be healed after the industry has closed, but we also wish that there were a few places where the old lake beds could be visited and seen.

Lake Albany eventually drained; its waters emptied out into the Atlantic Ocean. This may have been a catastrophic event, and we will return to this topic in chapter thirty-four. For the moment, however, we would like to have you see one place where the drainage of the old lake has left its mark. That would be the mouth of Roeliff Jansen Kill, where it empties into the Hudson River at Germantown. You can go and see this location from the bridge where Rte. 9G crosses the "Roe Jan." Here you will see towering cliffs of slate and shale, especially on the south side of the creek (fig. 10-2). The Roe Jan is a canyon here, and that canyon was carved by the creek during the millennia that followed the draining of Lake Albany. The floor of Lake Albany lies at about 190 feet in elevation

Figure 10-2. The mouth of the Roeliff Jansen Kill, west of the Rte. 9 Bridge.

in this region. The Hudson is at sea level here, so the Roe Jan was forced to carve its deep canyon in order to descend down to the level of the river. The Rte. 9G canyon speaks to us of a landscape recovering from the Ice Age.

11

The Great Deltas

The Hudson River today has numerous tributaries flowing into it. It is likely that most of them were present at the end of the Ice Age. But the Hudson itself did not exist back then; it was, as we have seen, flooded by Glacial Lake Albany. The lake was broad and it was deep. Rivers flowed into it, and they carried sediment with them. What happened to all that sediment? It was deposited into growing features known as **deltas**. Lake Albany is long gone, but many of its deltas are still there, right where they formed so long ago. They are ice age relics, and important ones in the Hudson Valley. These ancient deltas are easy to visit and very easy to see.

But first there are a few basics that you need to know about deltas. The classic view of a delta is roughly as follows: the rivers and creeks that produce deltas deposit sediments right up to the lake's water surface levels. When new sediment is then carried into the lake, it bypasses the older materials and is deposited farther along on the outer fringe of what soon becomes a growing delta. This, over time, creates a platform that has a flat top. That surface, again, lies at approximately the old lake water level. These flat-topped surface-level sediments are known as **topset** deposits.

Beyond the topset, there is a steep slope at the edge of the delta. Sediment carried by the stream flow falls over this edge of the topset and tumbles toward the bottom. A lot of the sediment is deposited on this slope and it has a name—it is the **foreset**. Foreset strata are typically inclined, often steeply (fig. 11-1). Below the foreset are the flat-lying strata of the **bottomset**. This is composed of sediment that made the complete journey past the delta and ended up on the lake bottom. There is, in fact, little difference between the bottomset and the lake bottom. Almost all of these delta deposits consist of fine-grained sands with a lot of mud. This simple topset, foreset, bottomset model goes back to the nineteenth century, and it is a bit out-of-date; modern views are more complex. But this view should do for our purposes, and it does, apparently, do a good job of describing lake deltas such as we see in the Hudson Valley.

Deltas continue to be deposited as long as there is a lake. But all lakes are doomed; they are condemned to eventually drain, dry up, or be filled in with sediment. None of them lasts forever. We have seen that Glacial Lake Albany came to be emptied as its water poured into the

Figure 11-1. Foreset deposits of an ice age delta in Manorkill.

Atlantic. As the waters drained away, the old deltas were left behind—no longer as lake features, but now as dry landscape features. These are called **hanging deltas**, and they are common and widespread up and down the Hudson Valley (fig. 11-7). Let's find out what to look for.

Any hanging delta is identified by its geomorphology. Each delta typically has a flat top; that, of course, is the old topset. Beyond that will be found the old foreset slopes. They may or may not display their original inclines. We will find, in chapters twenty-seven to twenty-nine, that the ancient foresets can be subject to landslide events that may alter them, sometimes considerably.

As we said earlier, the old rivers and creeks that produced the deltas commonly still exist. They flow out of the hills and across the topsets and down the fronts of the foresets. As they are typically erosive, they tend to carve canyons into both the topsets and foresets. When this is the case, the result is called a **cloven delta**. A vee-shaped canyon splits the old delta into two halves, just as the hoof of a cow is split, or cloven. Hanging and cloven deltas are both common in the Hudson Valley. Let's go and see.

What must be the finest hanging, and also cloven, delta in all of the Hudson Valley makes the platform that most of the town of Hyde Park sits upon. And the best place to go and see the Hyde Park Delta is at the Vanderbilt Mansion. The Vanderbilts picked the site for its scenic vantage point. Their house was constructed at the top edge of a foreset. From this location a sweeping vista of the Hudson Valley was obtained (fig. 11-2). Away from the river, the enormous front lawn of the Vanderbilt Mansion represents the topset of the old delta (fig. 11-3). It was, long ago in the nineteenth century, planted with scores of ornamental trees, all of which have long since reached their maturity. The delta topset and its most venerable trees make an unforgettable landscape. If you get the chance to walk about on this lawn, please remember that you are probably up to your knees in the waters of Lake Albany. Look at all the trees, and then imagine the lake waters. Rearranges your sense of reality, doesn't it?

But what about the cloven part of this cloven delta? When you enter the grounds of the Vanderbilt Estate, you must cross a stream with the

Figure 11-2. View from Vanderbilt Mansion looking downslope of the foreset of the Hyde Park Delta.

Figure 11-3. Topset of the Hyde Park Delta; also the lawn of the Vanderbilt Mansion.

Figure 11-4. The Crum Elbow Clove. Crum Elbow Creek cut through the Hyde Park Delta here.

unlikely name of Crum Elbow Creek. You drive across it on an attractive old bridge, and if you pause and look down you will see that the creek has carved a fairly sizable canyon (fig. 11-4). That is the clove. You can explore the estate and locate other views of the clove. It's a very good one.

There are quite a few other hanging deltas in the Hudson Valley. Another good one to visit is the delta at Elizaville. By taking County Rte. 19 north from Rte. 2, you are traveling upon the topset of the Elizaville Delta. There are two lakes on this topset, collectively called Twin Lakes. These, themselves, are ice age features. This delta was what is called an **ice cored delta**. When its sediments were being deposited, they buried two large masses of ice that had formed within the near shore reaches of Lake Albany. These two masses of ice melted away long ago and left holes in the ground, which are today filled with lake waters. Ice cored deltas are not uncommon, and this is a very good one.

But if you take Hapeman Road, off of Rte. 19, and travel around to the front of the Elizaville Delta you will see an unusually good display

Figure 11-5. The Elizaville Delta foreset.

Figure 11-6. Topset and foreset beds at Elizaville.

of its foreset (fig. 11-5). When we traveled there, we saw that quarrying activities had cut into the delta (fig.11-6). This exposed both the topset beds, which are seen as horizontal strata, and the foreset strata, which

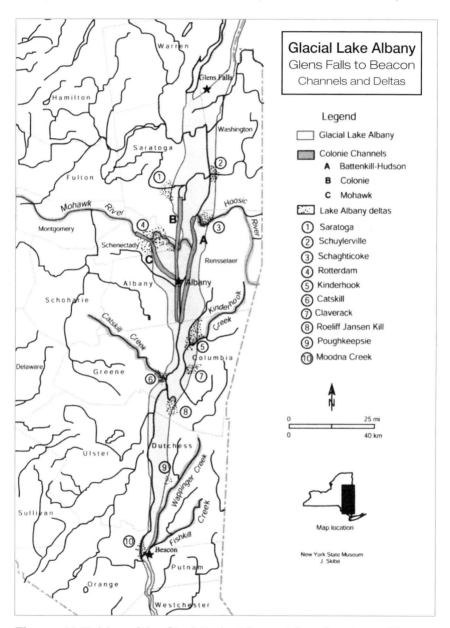

Figure 11-7. Map of the Glacial Lake Albany deltas. Courtesy of the New York State Museum.

are inclined. Below these we saw more horizontal strata—the bottomset beds. We don't remember ever having seen the structure of a delta as well exposed as it is here. See it soon—quarry cuts like this do not last very long; we hope this one will be the exception.

Still another hanging delta well worth the visit is the one at Mount Marion, west of the Hudson. This delta was turned into a housing development back during the 1950s. The well-drained topset made a perfect site for development. The delta's development lies adjacent to the Glasco Turnpike. This delta was produced by Plattekill Creek when it flowed into Lake Albany. It is a cloven delta, and the development lies upon the south half of the clove. Plattekill Creek first deposited the delta and later cut the clove. What is especially good about this delta is that there are a number of roads running through the development that lies atop it. You can drive around here and get another good sense of what the flat topset of a delta is like.

Space does not permit us to describe all of the other deltas of the Hudson Valley, but we have presented a map view of them (fig. 11-7). Each can be explored by anyone interested in doing so.

12

April 3rd, 16,190 BP

It's dawn on April 3rd, 16,190 BP. The sun is rising behind us on the eastern horizon of the Atlantic Ocean. We are 10,000 feet above the sea and just a bit offshore. To our north lies what must be a prehistoric Long Island coast. It's much as we would expect. There is a line of barrier islands; their quartz beaches are gleaming in the first light of the day. Behind the beaches are the coastal dune systems and the ragged, scrubby foliage that populates them. We rise a bit and see the bulk of this sandy island. Beyond is a broad lagoon; it must be an early Great South Bay.

We soar skyward to 20, 30, and 40 thousand feet in altitude. The pine forests of old Long Island are now spread out before us. They are a great unbroken expanse of conifers. In this time, long before human habitation and development, the forest is primeval—beautiful and dark green in the rising sun. But something is wrong; we can see clearly to the familiar sharp bend of the Connecticut River on the far northern horizon, but it is so far away. It must be some 150 miles to the north. Long Island, it would seem, is at least twice as wide as it should be. Now we realize that the ancient landscape below us must actually lie at what we know as the floor of the modern world's Atlantic Ocean. This sort of thing used to amaze us, but now we know better. That is the ice age landscape below; glaciers are large, and sea level is far lower than we have known in modern times.

We drop down to about 1,000 feet in altitude and continue north, paralleling this ancient coastline. Off to our left, in the far west, is another shoreline. It should be New Jersey, but we have reached it far too soon. Again there are just the right, bright quartz, barrier beaches. But something is wrong. The northern Jersey shore should have a fairly impressive ridge of low sandy hills—the coastal highlands—but these are not to be seen. And there is no Sandy Hook. We arrive at the coast a hundred miles or so east of where we expected it. Like Long Island, the Jersey coast seems to have greatly expanded seaward. But, again, we now know too much to be fooled. Once again we are looking down upon an ice age landscape that, in our modern times, will be seafloor. There is little escaping the obvious—the ocean waters are hundreds of feet lower than that which we, in our time, have known. The shorelines have expanded south and east.

We drift on through the sky. To the northwest is the mouth of a very large river. At least there can be no mistaking *this* geography; from its location, it must be the Hudson River. We are, quite naturally, drawn to this familiar feature. But, as we approach it, we see that it too is different. The Hudson flows through a canyon that looks to be maybe several hundred feet deep. Steep, gravelly slopes rise on either side. And the river itself is different. The lower Hudson has always been known as something of a serene estuary of the sea. That's not what we see here. It is an angry, powerful river in full flood. Swirling, churning waters flow swiftly toward the sea. The waters are brown with sediment; this Hudson River is far more erosive than the one we have known in our time. It has carved the deep canyon into the exposed sands of an emergent sea floor. This doesn't solve any old mysteries so much as it poses new ones. Now we are not so cocksure of ourselves. We drift northwestward, following the powerful stream, drawn toward the expected sight of the New York metropolitan area.

To the east, the low hills of northern Long Island finally appear and we regain our bearings. But now another mystery appears. Those familiar hills wind westward across Queens, Brooklyn and Staten Island, just the way they should, but there are no breaks between the three locations. However, we have already learned about this series of hills; it is the moraine of the Laurentide ice sheet. Between Staten Island and Brooklyn the hills extend right across the path of the modern Hudson River, and they have dammed the river. The sight is a wonder; a powerful flow of water turns west and passes through what will someday be called the Arthur Kill, which in modern times separates Staten Island from New Jersey. The loud flow roars on through the Arthur Kill and continues to the southeast as the strong flow we have just seen.

North of the great earthen dam, the New York City area is all different from anything we could have imagined. Most of it is simply under water. The moraine dam has created an enormous natural reservoir, a glacial lake. In our modern century people call this Glacial Lake Albany. We have heard of this lake, but now we actually get to see it, and we are awed. We drift slowly across what should be Manhattan Island. It's there, but it is so tiny; just a little land pokes above the waters in what will someday be midtown. All the rest is submerged by the lake. We drift to a near halt now. We half expect the arm of the Statue of Liberty and the tops of the many skyscrapers to rise above the water, but they are not there of course.

To the immediate west lies another surprise. The Palisades, that cliff of basalt that today towers above the Hudson, forms the western shoreline of Lake Albany. Beyond the Palisades is the broad, flat expanse of another large body of water. This is another great ice age lake, called Lake Hackensack. It is a big lake, much wider than Lake Albany. Its basin extends westward nearly to the first ridge of the Watchung Mountains. Beyond it lie the several ridges of New Jersey's Watchung Mountains, and then, within them, we see still another lake—Lake Passaic. The waters of this lake are trapped within the inner ridges of the Watchungs. Lake Hackensack is different; it was the product of that earthen dam, the moraine that blocked the Hudson River (fig. 7-1).

To the east of "Manhattan" lies still one more large lake basin, which is called Lake Flushing. This old lake covers much of what will one day become Queens and the Bronx. It, too, is the product of that same long earthen dam. The dam is a remarkable relic from an earlier moment in the Ice Age. All this would have been a mystery just a short time ago, but we have learned so much about the Ice Age. This moraine dam represents the southernmost advance of the Laurentide ice sheet. Here the glaciers had halted and deposited their gravelly sands in the enormous heaps that now make up the dam. This ice age wonder had made nearly all of the New York metropolitan area a land of large lakes. We know all this and we should be jaded, but we are seeing it for ourselves, and that is much different from just knowing about it.

We continue our journey up Lake Albany, past Manhattan to what is in modern times the Tappan Zee. Here, at least, things are just a bit more familiar; this stretch of the lake is but a deeper version of today's Tappan Zee. To its north we enter the Hudson Highlands. Here, again, is a deeper, but familiar, replica of today's Hudson Valley.

We continue north, drifting slowly above the lake. Because we are the mind's eye, we can do anything we want to, and now we want to go fast. We pick up speed and skim across the waters, 30 MPH, 40 MPH … 50 MPH. To the left and right, the banks of Lake Albany zip by us. There are still many forested areas, especially low along the banks of the lake. The trees are spruce and pine; few other species can be found. The ones we see are mostly small and young. As we continue north, the forests become fewer and patchier in their distribution. In between are bleak, gray landscapes. High up on the hills, snow fields appear and then become common.

We pick up even more speed, 60 MPH, 70 MPH … 80 MPH. Now the banks are rushing by in a blur. Trees become rare, and the landscape becomes bleaker and, if possible, grayer … 90 MPH. We are just above the water and can feel a bit of spray. Soon small cakes of ice appear—just a few at first. The blocks of ice soon become abundant and more closely spaced. It's only natural to slow down, 60 MPH … 40 MPH. The icebergs are very common now, and more of the blocks are large …

30 mph. The banks of Lake Albany are barren and desolate. We slow to a crawl.

Now, to our left and right, are great masses of **dead ice** glaciers. They are huge, immobile masses of stagnant, dirty, melting ice. They fill and clog the Hudson Valley as far east and west as we can see.

This trip has taken us past so many ice age marvels: the low sea level; the great earthen dam at Manhattan; the several lakes, especially Lake Albany; and now the dead ice. The world of 16,190 BP is so different. Great wonders can be difficult to fully appreciate, but not always. In our journey up Lake Albany we have been slowing down. Now, abruptly, we come to a halt and we will see what we had expected. A short distance to the north of us is an active glacier.

From as far to the northeast as we can see, and as far to the northwest as well, the Hudson Valley is filled with ice. An enormous glacier is actively advancing but also, at the same time, melting. Great wells of dirty gray water are burgeoning up from the bottom of the lake just in front of the ice. The glacier is a great, noisy, broken, cacophony of ice. The front of the ice looms high in the valley; it's broken into a chaos of jagged blocks by a myriad of complex crevasses. Periodically, masses of ice collapse into the lake and, as the great bergs well back up, huge tidal waves radiate down the lake. This is the source of all those blocks of ice that we have passed.

We rise hundreds of feet and drift northward, now quite slowly. The Hudson Valley glacier fills the valley and extends to the north as far as can be seen. It is broken by broad, curved, dark blue crevasses that betray its ongoing, slow, southward-creeping motion.

To the west, streams of ice have been diverted up Saw Kill Valley. We follow one. All of Woodstock, and Bearsville, and most of Overlook Mountain are submerged by the glacier. Ice is different from water; pushed from behind it can literally flow upstream. That is what we are seeing at Woodstock. Just to the north, two more tongues of ice streams flow up into the highest reaches of the Esopus Creek and Woodland Valleys. Branches of the ice press up the Stony Clove Creek and Bushnellville Creek Valleys and penetrate into the highest parts of the central escarpment of the Catskills.

To the north, thin mists only partially enshroud many of the high peaks of the Catskill Mountains. It is a magnificent sight. Bare rock is exposed at the top of each mountain. Just below, all of the upper cloves of these peaks are white with ice and snow. Before us, the entire eastern Catskills appear to be a great Alpine domain. Each mountain clove seems to have its own glacier within it. From each there descends a tongue of ice into the valley below. Sadly, however, the high mountain mist conceals the details.

We drift northward across those misty mountains. Beyond is West Kill Valley which, on this day, is hardly a valley at all. It is nearly entirely flooded by the ice of a glacier that is advancing up the valley toward the western slope of Hunter Mountain.

We continue onward, drawn to the northwest. Another fine tongue of ice has recently descended the upper Schoharie Creek Valley, and some of it is squeezing through the narrows of Grand Gorge. South of this, the front of the ice is melting, and a great swollen river of dirty meltwater gluts the Pepacton River south to Margaretville and then off to the west down the valley of the upper Delaware River. It is a strange image, but we are growing very used to strange visions.

We soar higher, two miles into the sky. Beyond Blenheim, a large valley glacier fills the Schoharie Creek Valley. To the east similar glaciers can be seen in Batavia Kill, in the upper reaches of Schoharie Creek at Hunter, and moving westward through Plattekill Clove. To the west another tongue of ice descends the upper Susquehanna Valley.

We rise up to fifty miles altitude. To the north another, even larger, river of ice moves westward through the Mohawk River Valley. To the west a great fan-shaped lobe of ice has planed across the Ontario Lowlands and approaches the Catskill region from that direction.

We continue our ascent. We turn eastward, and the whole extent of the Hudson Valley glacier can now be seen. It passes east of the Adirondacks, and it's a branch of this ice that gives rise to the Mohawk glacier. The Hudson Valley flow of ice continues to pass south beneath the Catskill Front until reaching its current terminus near Poughkeepsie.

From our perch on high, we can now begin to appreciate the full magnitude of this late phase of the Ice Age. It is truly breathtaking.

We soar still higher, to 100 miles altitude. An enormous lobe of ice reaches toward the western Adirondacks after completing a journey across most of eastern Canada. It extends all the way back to Labrador, from whence it formed.

Now we are at 1,000 miles in altitude. Another similar ice sheet extends north of the Ontario lowlands from deep into northern Canada. Virtually all of Canada is a nearly featureless plane of ice, an ice age replica of what can be seen in much of today's Antarctica.

We rise higher, to 15,000 miles above the Earth! Now we can see the whole northern hemisphere of the planet. Much of that hemisphere lies beneath a cap of ice. It all lies entirely below us now—the ice cap

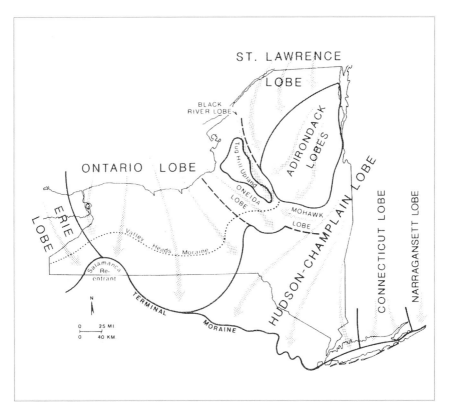

Figure 12-1. General flow of the ice across New York State. Courtesy of the New York State Museum.

covers almost all of Canada, and beyond that, it reaches across the Arctic Sea. It covers all of Scandinavia and much of Northern Europe.

Only now are all of our journey's remaining mysteries fully understood. As we have seen on our trek, the enormous sheet of ice has been melting. Its retreat has left behind all of the wonders we have seen. This ice age is a great event in the history of our planet. But on this April 3rd, 16,190 BP, the ice is in the midst of the long process of giving up its grip upon the earth, and it will eventually be nearly gone, melted almost entirely away ... *almost*.

13

The Glaciers of Plattekill Clove

We have been arguing in this book that there really was a Hudson Valley glacier, and earlier (chapter four) we cited evidence from Plattekill Clove. We found glacial striations that indicated a single substantial glacier had advanced down the Hudson Valley. With time, it grew and a branch of it swelled up into the highest reaches of Plattekill Clove. After that, a great ice sheet spread across the top of the clove, over all of the eastern escarpment and across all of the Catskills themselves. That, we noted, was the Laurentide ice sheet, which succeeded the Hudson Valley glacier. More striations indicated that the Plattekill Clove ice was gradually deflected southwestward as this ice sheet overran the whole region. The evidence from all these striations convinced us that the forward guard of all this was a Hudson Valley glacier. Our main point, again, was that there was a difference between the Hudson Valley glacier and the Laurentide ice sheet. These were two separate, but successive and intergrading, entities.

In this chapter we will travel forward through time and return to the Platte Clove vicinity to see activity associated with yet another, much later, advance of the Hudson Valley glacier. We will observe that once again a sizable tongue of ice peeled off from the main Hudson Valley glacier and then ascended Plattekill Clove. But this time, late in the Ice Age, it was a different sort of glacier and, more importantly, this time there was no Laurentide ice sheet.

The valley above Plattekill Clove extends westward six miles to Elka Park. It contains the uppermost reaches of Schoharie Creek, which flows off far to the northwest. This chapter is about this valley's late–ice age history, and there is much to relate. With experience geologists learn how to "read" the landscape. They learn how to decipher the geological history of any vicinity by finding out what to look for. We would like to take you down this valley, from the crest of Plattekill Clove westward to Elka Park, and give you some of that experience.

Figure 13-1. Valley, just west of the crest of Plattekill Clove. Here, valley walls crowd the stream.

The Glaciers of Plattekill Clove

Figure 13-2. Valley at Prediger Road. The road descends a gentle slope to the wide, flat bottom of the old ice age lake.

A careful observer of the landscape will find that there are two sorts of valley here. That may seem like a curious claim; after all, how could there be two types of valley? But, bear with us. These two valley types are different and alternate with each other through successive stretches of upper Schoharie Creek. The first type is initially encountered along County Rte. 16 at the crest of the hill that is also the very top of the clove. Here Plattekill Creek is squeezed between two embankments, one to the north and one to the south of both the road and the creek. The valley is narrow here and its walls close in on the stream (fig. 13-1). It is generally forested. Hikers enjoy this area, and a number of trails ascend the slopes of nearby Plattekill Mountain and High Peak. This stretch begins just west of the stone bridge at the Devil's Kitchen, and it extends west from there for about a mile.

But if you drive only a short distance farther to the west on Rte. 16, you will encounter the second, very different, type of valley. This begins at Prediger Road (fig. 13-2) and continues on to Dale Lane. We have entered the valley of Schoharie Creek, and we see something very different. The valley is not crowded, nor is it forested, so we can see its wide floor quite clearly. It is poorly drained, and perhaps that's why trees do not do well here.

We continue west another 1.3 miles and there the road ascends a slope, cresting near Farrell Road. South of the road, where we cannot see it, Schoharie Creek has again been funneled into a narrow valley crowded with steep banks on either side. This narrow stretch mirrors what we saw east of Prediger Road; it's very different than the terrain between Prediger and Dale Roads. We have returned to that first type of valley.

The creek seems to have struggled to cut its way through this stretch. It has. Schoharie Creek has been forced to erode a path through a geological feature, a heap of earth. This is the likes of something that we have seen before. This is a moraine—actually two of them, counting the other one back east of Prediger Road. Now it is time for us to come to understand what we have been seeing.

Earlier, we saw very much larger moraines in the Hudson Valley and especially at Long Island. We found that a moraine is a heap of very coarse-grained sediment. It is composed of sand and gravel, cobbles and boulders—sometimes very large boulders (fig. 5-2). This deposit was bulldozed to where we see it today by the advance of a glacier. The moraines of Long Island were the product of the Laurentide ice sheet, and they are very large. The moraines up and down the Hudson Valley were produced during the retreat of the Hudson Valley glacier. Those moraines are a lot smaller. The moraines of Plattekill and Schoharie Creeks were produced by something much smaller still. It was a relatively small valley glacier that did the job all along this, the upper reach of Schoharie Creek. The Schoharie Creek Valley here was a scale model of the Hudson Valley, and it had a scale model of the Hudson Valley glacier in it. A long time ago, perhaps 18,000 years, a glacier expanded up out of the Hudson Valley westward through Plattekill Clove and onward as a flow of ice that penetrated the upper Schoharie Creek Valley. It advanced all of the six miles that we have been driving.

This was a streamlike mass of ice; we sometimes describe it as a "tongue" of ice. The glacier filled the upper Schoharie Valley, quite possibly to near its top. This could happen because there was no Laurentide ice sheet to deflect the ice to the south as had happened earlier. We are not sure just how far to the west it advanced, but the advance seems to have reached at least a temporary halt at the Farrell Road site. At this stop the glacier would deposit its moraine sediments. Time would pass and, eventually, the climate would warm and the glacier would begin to melt back eastward from here. It was the melting of the ice that accomplished the retreat. If we had time-lapse photography of the event, we would see vast quantities of water pouring off the ice as it retreated. That ice water flowed westward into the Schoharie Creek. If you drive on to Elka Park Road and turn left, you will encounter a very flat landscape just left (east) of the road. This appears to be an outwash deposit, associated with the Farrell Road moraine.

The retreat of a glacier was always a complex affair. In the upper Schoharie Creek Valley it was a small-scale version of what we have seen back in the Hudson Valley (fig. 8-1). Retreat was not smooth; it was interrupted by chapters when the climate grew colder again and the ice managed brief advances. As we have seen before, it was a two-steps-backward-and-one-step-forward sort of process. The glaciers retreated a lot, advanced a little, and then retreated a lot once again, as the climate fluctuated between warmer and colder times.

Each time the glacier advanced it carried coarse moraine sediments forward. Each time it retreated it left a heap of earth, a new moraine that dammed the flow of the Schoharie Creek. There are two such moraines that we have seen: one extends west from the Devil's Kitchen; the other is found at Farrell Road. The damming effect of these produced at least one glacial lake between the two moraines, extending from Prediger Road to west of Dale Road.

And now we understand that second category of valley here. The broad, flat-bottomed stretch of valley represents the floor of an old glacial lake that once occupied a sizable stretch of this valley. Turn south onto Prediger Road and you are descending the slopes of the old lake down to its bottom. Turn south at Dale Road and you do the same, but Dale Road also takes you across the entire width of the lake, a half mile. This lake must have been about thirty feet deep; that's the change in elevation from the lake's Rte. 16 shoreline to its valley floor bottom.

Now, any time we drive back to the east toward Plattekill Clove from Elka Park, we travel with a lot of new knowledge. We drive past Farrell Road and we see that we are on a moraine; here the valley is clogged with sediment. We drive past Dale Lane and Prediger Road and we see an ice age lake; the valley floor is today broad and flat, with meadows. We can pause and, in our mind's eye, gaze out across this old lake. All around its edges we see platforms of floating ice. The center of the lake is dark and still. There is just the first stage of the return of forests in the hills rising above the lake. A few small spruce and pine trees are starting to grow. If we are very lucky we might even see a mastodon walking along the shore.

But what it didn't take was luck to see this; it just took knowledge and understanding of the ice age heritage of this pretty valley above Plattekill Clove.

14

Ice Age Water Park

Generations of hardworking glacial geologists have carefully done the fieldwork that has documented the numerous positions that mark stages in the retreat of the Hudson Valley ice (fig. 8-1). This deglaciation was a very complex affair. We are forced to generalize their work in order to fit our purposes with this book. Still, there are some locations where it would be a crime not to attempt to flesh out all the details, and we feel that way about the region around North Point, the high mountain at North/South Lake State Park. As we studied the area we encountered the remarkable late–ice age equivalent of a water park.

It all started years ago with a climb up to the Rip's Rock ledge. That's the wonderful horizon of sandstone that towers above the old site of the Rip Van Winkle House in (what else?) Rip Van Winkle Hollow. You can see the ledge from most vantage points on the edge of the escarpment in the park. It overlooks the Old Mountain Turnpike, the nineteenth-century carriageway that once carried guests to the Catskill Mountain House hotel. Nowadays the old highway is just a dirt hiking trail. The trek toward Rip's Rock ledge starts on that trail. It then follows the streambed up the canyon above the bridge at a sharp bend in the old trail. Go about 100 yards up the canyon, then turn right (north) and ascend the steep slope. You will soon find yourself funneled into an increasingly steep and narrow, dry canyon. At the very top, two vertical rock cliffs form the box canyon of one of the finest **glacial spillways** we have ever seen (fig. 14-1).

It's really a gem. Long ago, near the end of the Ice Age, the glacier still filled much of the Hudson Valley. It abutted the north side of the great mass of rock that is Rip's Rock, but it was melting. Geologists call this dead ice. It is not advancing; it is just lying there and melting away. Ice water cascaded off of the melting glacier and, as it was funneled past Rip's Rock ledge, it carved that spillway. The water continued down the canyon that is now Rip Van Winkle Hollow, and it is this flow of water that eroded the deep canyon that's down there right next to the marked hiking trail. Today, as in the past, waters in that canyon tumble down into the Hudson Valley. Like all such spillways, it is a wonder to ponder the terrible currents that so long ago passed this way. And it is also a pleasure to understand the origins of the ravine we find in the hollow.

Once we had discovered this spillway, our eyes became attuned to looking for more, and we met with success very quickly. We took another

Figure 14-1. The dry spillway canyon below Rip's Rock.

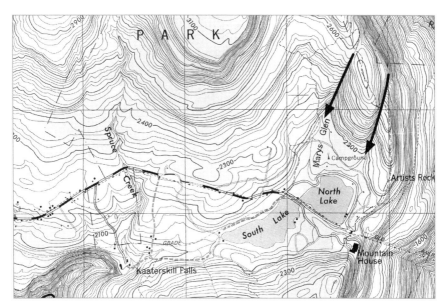

Figure 14-2. Map of spillways in North Lake region. Base map courtesy of U.S. Geological Survey.

look at the topographic map of the North Lake area. Its contour lines betrayed hidden landscape features, and we immediately saw two more spillways (fig. 14-2). The first one had cut its way through the Catskill Front at the top of Mary's Glen, at just less than 2,600 feet in elevation. It begins at the marsh a bit south of Badman Cave. This is a popular hiking destination on the blue trail below North Point. Once again this is a feature that records a moment at the end of the Ice Age. The first spillway carved a shallow but distinct canyon all the way to Mary's Glen and on to North Lake. The second was a similar canyon also carved into the Catskill Front, a quarter mile to the south, behind Sunset Rock. This one began at a little more than 2,400 feet in elevation. It seems to have at least helped carve the cliff there. It, too, could be followed down to North Lake.

Now we were caught up in the project. We pored over our map and saw, very much to our astonishment, that to the north, at Dutcher Pass, there was not just one spillway, but actually two of them. The second one lay immediately north and uphill from the first (fig. 14-3).

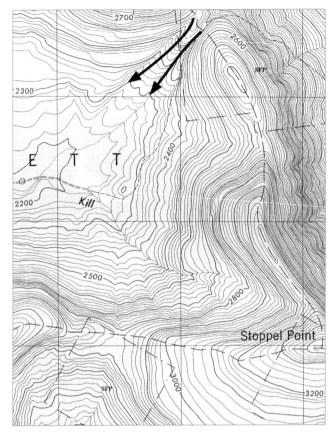

Figure 14-3. Close-up map view of spillways at Dutcher Pass at top of map. Base map courtesy of U.S. Geological Survey.

It's one thing to discover spillways on a map; it's quite something else to go and see them for yourself. We were itching to do this. The plan was very simple: first we would hike up the Mary's Glen canyon, turn south on the blue trail and then descend the lower canyon. The Mary's Glen canyon turned out to be pretty much what we had expected. It was wide in its lower reaches and then quickly narrowed uphill. In our mind's eye we could imagine the pounding white-water rush that had once passed this way. When we reached the top, we were on the blue trail at the peat swamp that is not too far south of Badman Cave. Here the spillway had cut its way through the Catskill Front. This one is not very sharply cut into the landscape, and it may not have been a very great flow of meltwater. But it did help to carve Mary's Glen.

The lower canyon turned out to be much more interesting. We found our way down the blue trail until we reached the turnoff to the yellow trail. That marked trail leads eastward to Sunset Rock, but we looked to the west and headed that way. There is no trail here, marked or otherwise; it's a tough trek, but it is worth the effort. Very quickly we found the upper reaches of that second spillway; let's call it the Sunset Rock spillway (fig 14-4). At first it was not very deep, but that soon changed. We discovered that this meltwater flow had been very powerful and very erosive. Soon a fine cliff appeared to the right, and then a deep gash opened up. This had been quite a chasm back when meltwater flowed through here. But today the bottom of this ravine is filled in with earth and dead and badly decayed trees. It bears no resemblance at all to what must have been here long ago. We were surprised at how little evidence there was of hikers having ever been here. It would seem that almost no one had passed this way for a very long time, if ever. This canyon seemed to have remained almost unknown in the two centuries that people have been exploring the region. At popular North Lake, that's incredible.

Figure 14-4. The lower, dry spillway canyon near Sunset Rock.

But our discoveries were going to get even better. All of these features are fine examples of what are called **paleoforms**. They are landscape features that date back to the conditions of ancient ice age times. About two-thirds of the way down this canyon we found something that is a very special paleoform. There, at about 2,300 feet in elevation, was an empty waterfall. We had discovered a dry cascade. It was now all mossy and overgrown with ferns, but we could easily imagine the thundering falls that once tumbled down it. At its bottom was a fine example of a **plunge pool**. Plunge pools are carved by the intense flows of falling water crashing down on the bedrock below the falls. Once, this one had been deep beneath a pounding curtain of cold glacial meltwater. Today there is no water, and the old pool has largely filled in with soil and vegetation. But, though now derelict, this wonderful fossil waterfall was the highlight of the day, and soon we emerged onto the road at the bottom of the canyon.

Our later journey to Dutcher Pass was also a fine success. The main spillway, the lower of the two, had been the path of the historic road that once crossed through the pass (fig. 14-5). That may be why it is flat-bottomed; it may have been "improved" to allow easy passage of traffic. We climbed north a very short distance on the blue trail and there, just as we had expected, lay the second spillway canyon (fig. 14-6). It was a beauty. Like the Sunset Rock spillway, this one showed no evidence of any foot traffic; hardly anybody has ever clambered down this canyon. So it, too, was in pristine condition; it had suffered few if any changes since the Ice Age. We found it a wonderment to climb down it and to once again imagine the ice age heritage that was here.

After exploring our newly recognized spillways, we had to assume once again the roles of proper scientists. The chronology of all this is of some importance. The spillways certainly formed in several stages during the melting of the dead ice. It is quite likely that the Dutcher Pass and Mary's Glen spillways formed during the earliest stages. They are the highest of the group, all three being above the 2,500-foot level. We can imagine being back at a time when the melting Hudson Valley glacier was lapping up against the very top of the Catskill Front in this region.

Figure 14-5. The lower of the two Dutcher Pass dry spillways,

Figure 14-6. The upper of the two Dutcher Pass dry spillways.

Later, as the surface of the ice melted down to about the level of 2,400 feet, the second stage of spillway formation began. The Sunset Rock flow would have been active for a period of time. Down below, the entire North Lake basin, however, seems to have been still filled in with ice. It was, we imagine, a warm glacier, something we call a **temperate glacier**. There was a lot of melting going on and there were, we suspect, a number of sub-glacial streams in the North Lake basin. We can imagine a stream flowing beneath the ice from the Sunset Rock spillway, across today's North and South Lakes, to the present-day location of Kaaterskill Falls, and from there on through an ice tunnel into a still-glacier-clogged Kaaterskill Clove. Such things, sadly, can only be imagined, not proven.

The Rip Van Winkle spillway came next. At 1,860 feet, it records the time when the Hudson Valley glacier had retreated farther to the north while being melted down to a surface about 600 feet lower than at Mary's Glen time. This must have been a very fine spillway, as Rip Van Winkle Hollow shows the effects of erosion by a very substantial rush of water. Once again it is easy to imagine this flow plummeting into a dark icy tunnel as it tumbled into the lower Hudson Valley.

We had pretty much worked out all of the above when we were greatly surprised to find yet another spillway. We were with the Mountain Top Historical Society, climbing Cairo Round Top, when the hike leader, Bob Gildersleeve, who was familiar with what we had been investigating, told us we would soon be seeing one. And there it was—a gem of a spillway (fig. 14-7). Here the ice water had cut a canyon about 100 feet deep. This one is clearly the youngest of them all. It is only at 890 feet in elevation, and it also lies northeast of the others. It has to be that the Hudson Valley glacier had melted back a great deal by the time of its formation.

Have you ever been to a water park? Perhaps you have been to Zoom Flume in East Durham, Greene County. If so, then you have seen a number of fine water slides. People ride them and have a lot of fun doing so. We had been doing the same, but our flumes were perhaps 15,000 years old and we did not actually ride them. In the end, you can imagine how much real fun it was for us to work out the history of this late–ice age water park. This was genuine scientific research, but this kind of field study can hardly be called "work." It's all good outdoors recreation, punctuated by the near glee that comes with those moments of discovery. The lives of geologists are like that—on good days.

This story focuses us on the remarkable end of the geologic history seen here. Within a two-mile radius of North Lake, there are four lovely

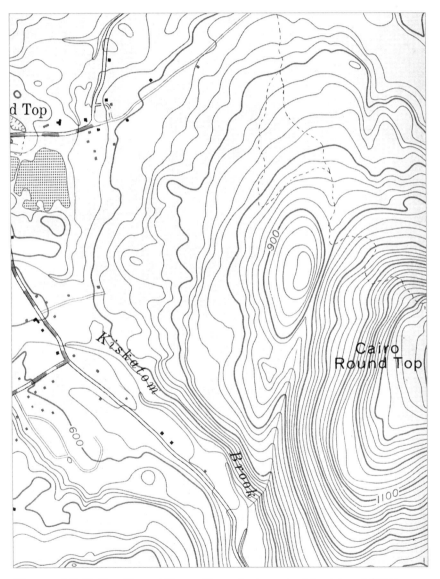

Figure 14-7. The Round Top spillway, just west of the top of Cairo Round Top. Base map courtesy of U.S. Geological Survey.

glacial spillways, and within four miles there are three more, for a total of seven. That's a lot of melting glaciers and a lot of meltwater erosion. This wonderful area is steeped in ice age history. North Lake positively astonishes us with its geological richness.

When we found the time, we climbed up to Boulder Rock on South Mountain. There, stretching out before us, was the whole Hudson Valley, but for us it was a late–ice age Hudson Valley. Ice filled the valley right up to the level of the Boulder Rock ledge. We looked out at a great dead ice glacier. It was a rainy, overcast day, and even with a huge glacier right before us, it was warm. The ice was gray and very dirty. It had been melting for quite some time, and mud had been accumulating on its surface. The rain was washing some of this away. Pools of dirty water were commonplace. Some of that water escaped and followed shallow brooks that led to holes in the ice. Swirling vortices of water disappeared down those holes. We listened to the gurgling sound that resulted.

The rain picked up, and large amounts of the warm rainwater started turning into a ground fog that rose up and enshrouded the ice. Our last impression was that the sound of the flowing water and its gurgling took on an echoing tone. All became white, the color of dense fog.

15

The Red Chasm

Kaaterskill Clove is one of the region's scenic centerpieces. Almost all of us, at one time or another, have driven up the clove highway, Rte. 23A. Many of us have visited Kaaterskill Falls and then hiked east along the trails that hug the cliffs that edge the north side of the clove. But there is another location, less well known. It is a small gorge within the big gorge, easily overlooked but very picturesque. We are speaking of that stretch of Kaaterskill Clove that runs for about half a mile downstream from Fawn's Leap, through the lowest part of the canyon. It's sometimes known as the Red Chasm (fig. 15-1). There, Kaaterskill Creek has carved a particularly narrow and scenic gorge, one certainly worth visiting and one very much worth understanding.

There are two bridges located at the downstream end of the clove; the first and lower bridge has no name as far as we know. The second has been called Moore's, or More's, Bridge ever since the nineteenth century. The Red Chasm is the section of Kaaterskill Creek that lies between those two bridges.

It's called the Red Chasm for two reasons—it's a chasm, and it is red! It's red because of its ancient geological history. The walls are lined with

Figure 15-1.
The Red Chasm below More's Bridge.

strata of soft red shale and red sandstone that date back to the Devonian Period, about 375 million years ago. These red sediments accumulated on something called the Catskill Delta, a great mass of sediment that eroded out of and became deposited at the bottom of a huge New England mountain range called the Acadian Mountains. Those mountains were ancestors of today's northern Appalachians, but that is another story. The "chasm" part of the name is aptly chosen. Kaaterskill Creek becomes quite narrow below Fawn's Leap, and there is one short stretch where the creek seems to have knifed its way into the landscape (fig. 15-2).

Figure 15-2. The narrows of the Red Chasm.

The first question a geologist might ask is, why is such a gorge here? It might seem logical that Kaaterskill Creek would have done most of its canyon carving farther upstream and that, by the time its waters had reached the bottom, their erosive energies might well have been expended. Well, that is not precisely the way it works; the stream still had plenty of erosive power left in it, even at the very bottom of the canyon. And here the flow encountered that thick sequence of red strata. These are, by rock standards, mostly very soft, and the currents had cut into them like a hot knife into warm butter, creating an exceptionally steep and deep cut.

There are a few more-solid strata down here as well, and they have made their contributions to the scenery. Just above the lower bridge is a fine thick horizon of tough sandstone. The creek had limited luck in cutting through it, and the waterfalls that resulted are quite rugged. Similarly, upstream, there is a second sandstone ledge and a second falls. Then there is the most famous falls—Fawn's Leap.

But, when we visited the clove, we saw that there was more to be discovered here. We ascended the chasm from its bottom. That involved some considerable wading and some climbing over falls. It was nice to see this scenic locality, but we felt that we needed to learn more. We remained curious about all this and studied a topographic map of the Palenville area. We found something odd there. Palenville, just downstream from the Red Chasm, is built atop some very curious landscape. Take a look at the map (fig. 15-3) and notice how the contour lines define what looks like the shape of an old-fashioned lady's fan. When we saw

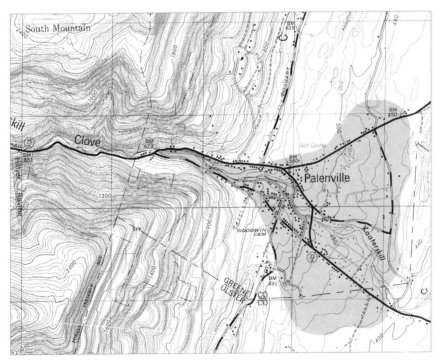

Figure 15-3. Topographic map of the Palenville alluvial fan. Shaded area defines the fan shape. Base map courtesy of U.S. Geological Survey.

Figure 15-4. Kaaterskill Clove from the air, with the alluvial fan and Palenville in the foreground.

that, we instantly recognized it as not a lady's fan, but what geologists call an **alluvial fan**. This is a heap of sediment that had once been carried out of the hills above and deposited where Palenville is located today.

You might notice this the next time you are driving through Palenville. When you head east on Rte. 23A, you will find yourself descending a long, gentle, downward slope that doesn't stop until you reach the Kiskatom Flats just east of Rte. 32. That's the fan, and practically all of Palenville is built atop it. If you have the good fortune to fly over Palenville, you can see the entire fan in one grand sweep. You will see that the roads of Palenville spread out very much as the old streams must have (fig. 15-4).

We wanted to take a good look at these deposits, so we hiked down the canyon from Moore's Bridge. Pretty soon we found a place where the river had cut into the banks and exposed the alluvial fan sediments (fig. 15-5). These were made of sedimentary grains of all sizes, ranging from boulders and gravel down to sand and clay. That's typical of an alluvial fan.

Figure 15-5. Exposed alluvial fan deposits along the slopes of the Red Chasm.

But there was still more that needed to be understood here. We felt that there had to be a relationship between the fan and the Red Chasm, especially that most remarkably narrow and very steep part, the narrowest part of the canyon. We wondered how exactly had all this wonderful scenery come to be formed. Actually, the solution was not all that difficult. We just had to think about what had happened here.

The building of the Palenville fan was only the first stage. A swollen late–ice age Kaaterskill Creek had carried enormous amounts of sediment from above and spread it all out into a fan-shaped deposit. There was no Lake Albany here, so topsets and foresets did not develop. Instead the sediments of the fan simply formed gentle slopes. The fan kept enlarging as long as there was a very plentiful source of sediment from the mountains above. But that would not be forever. Once the Ice Age was over, reforestation eventually occurred. As the trees took root, they stabilized the landscape and slowed down the rates of erosion. That shut off the source of much of the sediment. That means that the growth phase of the Palenville fan had only been a relatively short period of time.

Once large amounts of sediment stopped coming down the canyon,

Kaaterskill Creek changed its personality. It stopped depositing sediment and switched to eroding it. This younger persona of Kaaterskill Creek began cutting into its own alluvial fan. It was easy to cut through those soft sediments, and Kaaterskill Creek must have quickly made a fine ravine here. When that creek reached the soft shales below, it continued to slice right through them. The shales are soft enough to cut, but hard enough to make vertical walls, and that's how the steepest, narrowest part of the clove formed. It was a simple solution to an interesting geological problem. This is a cloven fan, and we have seen things like it before, especially at the Hyde Park Delta.

After our work in Kaaterskill Clove, we traveled south to Plattekill Clove, and there we saw a virtual repeat performance of all this. Here, too, was another, although smaller, alluvial fan. The village of West Saugerties lies upon it (fig. 15-6). Once again we could imagine vast volumes of foaming white water cascading down another great canyon. It eroded, widened, and deepened that clove, exactly as we had seen in Kaaterskill. We had gained knowledge at Kaaterskill Clove and had brought it along with us on this visit to Plattekill, and that had equipped us to look at the landscape here and to understand it.

We thought that we understood Kaaterskill Clove now, and we made one more visit. We hiked up to the top of Fawn's Leap and gazed down the canyon. Below us, several teen-aged boys splashed in the waters, raising quite a ruckus. It was summer, and Kaaterskill Clove was beautiful. We had learned a lot here, and that had taken us back to a time just after the Ice Age …

For us it was another summer, nearly 14,000 years before our own time. Down here the Ice Age was over; it was sunny and almost hot. It had been warming up for quite some time. In the nearby slopes we could see a number of young pine and spruce trees. Nature was reclaiming what the Ice Age had recently given up. But high above, within the upper clove behind us, it was still the Ice Age. Up there, in the shadows of the mountains, it was still cold and there was a remnant "cloveful" of melting dead ice.

Figure 15-6. Plattekill Clove from the air, with a small alluvial fan in front of it.

The Fawn's Leap site was deafeningly loud; enormous volumes of glacial meltwater were cascading past us and on down Kaaterskill Creek. The roar was echoing off of the high cliffs. The flow streamed across the Fawn's Leap ledge and continued eastward in a powerful rush. In this time, Kaaterskill Creek behaved like an enormous fire hose. We observed several rainbows, spaced at intervals in the canyon below.

The torrents of Kaaterskill Creek were laden with sediment that the waters had eroded out of the highlands above. There was little trouble carrying the heavy weight of this sediment as long as the water was flowing so rapidly out of the mountains, but when the currents reached the bottom of the canyon they lost momentum and slowed down. The weight of the sediment then became more than the slowing currents could bear, and they began depositing their burden in the great heap that glutted the lower canyon.

Below us, the wet sediments were bright red and dotted with thousands of cobbles and boulders. To the left and right of the clove, the sediments abutted on the vertical red walls of the canyon. They nearly, but not quite, filled this lower stretch of the canyon. There have been, throughout time, many floods in Kaaterskill Clove, but this is the only time in history that the clove was flooded with sediment.

Flows of water were entering the lower clove from several other directions. From the north side there was the pounding rush of a meltwater cascade descending from just west of Palenville Overlook. Across the canyon and high up on the south side, at what is today called Hillyer Ravine, was a small **Alpine glacier**. Its ice had accumulated beneath Poet's Ledge and hung from those rocks like a great icicle. It was melting, and here another thundering curtain of water plummeted down into the clove from up above.

Beyond, and downstream quite a distance, the clove widened and opened up. So did the alluvial fan. Upon it, the flow of Kaaterskill Creek split up into many shallow, crisscrossing channels. We have seen these before (fig. 1-1); geologists called such waters "braided streams." These meandered across a widening mass of wet, glistening sediment. This was the great Palenville alluvial fan itself, and we were witnessing it in its prime. It was a bleak landscape, but not entirely sterile. Poking out of the muds were the tops of numerous small pines and spruce. The poor trees had had the misfortune of taking root in these fresh sediments. They had hardly started growing when they began to be buried in the rapidly accumulating sediments. Their struggles for life would be futile; they would soon be completely buried, so rapid was the deposition rate here. Beyond, and a bit farther east, it was even worse; the front of the expanding fan was advancing, and its sediments were expanding eastward, burying a whole forest of spruce and pine in the process. Nature was building a fan and killing a forest—nature didn't care; she never does.

In the far eastern distance we saw the blue of a sizable lake. Here the meltwater flow of the post-glacial Kaaterskill Creek was impounded by a temporary dam. The meeting of the warm air and cold waters generated a thick fog that formed and drifted across the surface of that lake.

When the fog lifted, it was once again the summer of 2009 and our dreamlike visions of the past were interrupted by the noises of today. Below us a middle-aged man had fallen, fully clothed, into the pool that lies under Fawn's Leap. His friends were chanting something silly about "Larry's Leap." We laughed. We had seen lower Kaaterskill Clove in all of today's natural beauty. To be there on a nice summer day is one of nature's pleasures. But to see it as it was 14,000 years ago, at the end of a great ice age, is one of geology's privileges.

16

The Bottom of a Lake

When you have been a geologist for a long time, you develop a real sense for the landscape; you gain insights, and you just plain notice things that most others don't. It gets better with age and experience, and that's important, as you become an increasingly effective observer in the field. In many sciences it is the very young who do the best work. In this science, however, the most seasoned eye often sees the most and the best.

There is one thing, though, that even the most experienced eye finds difficult, and that is seeing what is not there. That happens when nature has painted a landscape but left something out. If you can notice that absence, you may be awakened to some wonderful moment in the geological past. But, just how do you see what is not there? Well, as we said, it comes with age and experience.

Let's pick up where the last chapter left off. Let's go to the intersection of State Rtes. 32 and 23A, not far from Palenville (fig. 16-1). Many of you have probably been by there many times. If you do go there, drive north on Rte. 32 and then turn east on Cauterskill Road. Along the way try to see what you don't see. You will have passed a fine flat agricultural field

Figure 16-1. Intersection of Rtes. 32 and 23A. This flat surface is the floor of Glacial Lake Kiskatom.

and a company that produces special eggs for lab work, but what else can you see? The answer is: well, not much. There are no canyons, no rivers, no hills, and no dales. In fact, there is just about nothing there. It's just a nice big flat field. We have traveled by there many a time, and we have long noticed that we weren't noticing much in the way of real landscape, just a lot of flatness. There is something else missing there, and it takes a lot of looking to see it. There are very few rocks lying about here. There are almost no cobbles and certainly no boulders. That's unusual in the American Northeast, where there are generally "two rocks for every dirt." We finally looked it all up in the geological literature and confirmed what we had suspected. This was the bottom of an ice age lake.

Turn right from Cauterskill Road to Paul Saxe Road and head south. To your right is the lake bed; you are driving along the old shoreline. Go past Mountain Turnpike and continue south on Ramsey School Road. Turn right on Rte. 23A and go west. You will return to Rte. 32 and will have circumnavigated most of the old lake. This broad flat landscape, so well suited for the farmer's plow, is sometimes known as the Kiskatom Flats. The flats mark the floor of an ice age lake named, naturally, Glacial Lake Kiskatom.

The story of this lake takes us back to where we left off in the last chapter. It was about 14,000 years ago when warming climates were bringing the great Wisconsin stage of the Ice Age to a fitful end. All of the Hudson Valley had spent most of the last 10,000 years under thousands of feet of ice, but that was nearing an end. At that moment the Kiskatom Flats were something of a glacial battlefield. The ice had earlier retreated halfway to Albany. Then the climate cooled briefly and the ice readvanced to what is now the southern end of the Kiskatom Flats. That readvance was temporary, and the ice was soon melting away, this time for good. But glaciers are dirty objects, and this one would leave behind very large heaps of earth all along what had been its southern edge. That's just a short distance south of the intersection of Routes 32 and 23A. We have learned about this sort of thing earlier; these deposits make up a small moraine.

Figure 16-2. Glacial Lake Kiskatom as seen from the Mountain House ledge.

As the ice melted away from the area, a landscape basin was left behind just north of those heaps of earth. With all the meltwater that is produced by retreating ice, this basin filled up quickly—hence the origin of Lake Kiskatom. It didn't hurt that, as we have seen, a swollen Kaaterskill Creek was flowing into the growing lake. The waters of Lake Kiskatom rose to an elevation of about 360 feet, and our estimation is that the lake was five miles long, north to south, and one mile wide, east to west (fig. 16-2).

It must have been quite a sight. On its northern shore there was still a great glacier, rising perhaps a few hundred feet above the lake waters. All along the eastern shore there may have been more, equally thick, glacial ice. This was, of course, the end of the Ice Age, and the climate was getting quite warm. All of the ice was actively melting, and vast volumes of meltwater were pouring out of the glacier. We already "witnessed" more water pouring into the lake from Kaaterskill Clove. Imagine thundering cascades of foaming gray water pouring into the lake.

We stand here and see, in our mind's eye, a glacier that was not melting so much as it was disintegrating. From time to time, enormous masses of ice would have detached and crashed down into the lake. Believe it or not, huge tidal waves would soon have rippled back and forth across the little lake—such things do occur in small lakes. The surface would have been dotted with numerous small icebergs from previous, similar collapses. None of these would have sunk the *Titanic*, but they were big enough to add a sullen but picturesque note.

What about that flat landscape? Well, we have seen these before when we visited Glacial Lake Albany (chapter ten). These are common landscapes throughout the Hudson Valley and also in the valleys of the Catskills. We are always on the lookout for them. It has been our experience that these almost always mark the locations of old glacial lakes. What happens is that the meltwater is dirty with sediment that quickly accumulates as flat, stratified sheets on the floors of the lakes. Much later, after all of the water has drained away, the flat lake bottoms become wetlands that slowly dry out into flat fields. The northeast corner of Kiskatom Flats is still a wetland. You can see this from Rte. 32 heading north from the Rte. 23A intersection.

There are similar flats west of the town of Woodstock. Those make up Lake Woodstock. There is another large, flat area just west of Cooper Lake; that's Glacial Lake Cooper. West of Windham there are old glacial lake flats. And there are more—you can start watching for them in your travels. It shouldn't take long before you find some, and with time you are likely to find a lot of them. This is something you will become accustomed to noticing.

But flats are not all you can see in the Kiskatom area. Lakes have to drain, and Lake Kiskatom was no exception. The water drained off to

Figure 16-3. The High Falls spillway once drained Lake Kiskatom.

the east and carved a fine spillway. Take High Falls Road almost three miles east of Rte. 32 until you find your way to the bridge over Kaaterskill Creek. The channel there is composed of bedrock, and it was carved at the end of the Ice Age when all sorts of water passed out of Lake Kiskatom and flowed off toward the Hudson (fig. 16-3). Take a good look at this channel and imagine the flow of water that was needed to carve it. If you can perform that act of imagination, then you will have a good idea of just what the end of the Ice Age was like here.

Having just spent five minutes reading this chapter, you are now more experienced and have a better-trained eye. It's time for you to start noticing such things as these. Please do go to the Kiskatom Flats and "see" the lake with its little icebergs, look at the glaciers to the north and east, and watch the pounding cascades of water. You are seeing what is not there, and that makes you a bit of a geologist.

17

Land of Lakes

Our quest to understand the late-ice age history of the Hudson Valley eventually led us to Taghkanic State Park in Columbia County. It's always a nice place to visit; you can enter the park all year round, and in fact they even have a winter trail there—but summer is when it is at its best. There are beaches, boat rentals, and hiking trails, or you can just go and picnic. Our notion of recreation, however, involved the Ice Age. We were drawn to Lake Taghkanic.

We visited with many questions in mind. The first was, why is the lake there? That might seem at first to be a silly question; after all, lakes are *just there*, aren't they? Well, geologists do ask such questions, and the answers often carry us off on journeys into the distant past.

It's not that we didn't already have some ideas about the answers, but we wanted to work on the problem and nail it down. Here's what we had guessed: Taghkanic, like most lakes in the northern reaches of the United States, is probably glacial in origin. If you have a good road map of Pennsylvania, you can see this for yourself. You will notice that lakes are very common throughout the northeastern part of the state. But, just south of an imaginary line, they become very rare. That imaginary line marks the southern reach of the glaciers at about 21,000 years ago after they had expanded to an ice age peak. North of that line is a land of lakes; there the drainage had been disrupted by the grinding passage of the ice. When the ice retreated, the glacially disrupted drainage produced those abundant lakes.

Well, what works in Pennsylvania probably works in the Hudson Valley, so we had come to Taghkanic Lake to see. We soon found a lot of exposed bedrock. Those outcroppings often had a planed-off appearance to them. That, we supposed, was the work of the ice (see chapter three about the Shawangunks). We were hoping to find striations and gouges in the park's outcrops, but we had no such luck. The local bedrock there is pretty soft stuff, by rock standards, and any striations that were there have been weathered away long ago.

But still, the outcrops do have that planed-off appearance. We think that the park was scoured by the passing of the ice, and that the Taghkanic Lake basin was formed that way too. The slowly moving glacier scooped out enough soft rock to create the lake. That happened often during the Ice Age. We have already seen such a thing at North Lake.

We kept hiking, and soon we found a lot more. At the southwest corner of the lake, there is an inconspicuous swampy outlet where water drains out of the lake (fig. 17-1). The flow passes through a narrow bedrock notch; steep rock walls rise well up above the stream. That's something we have seen a lot in the Catskill Mountains (chapter fourteen), but never before down here in the valley. There was a story here.

This canyon is a glacial spillway; it takes us back to the very end of the Ice Age, when the glaciers were actively and rapidly melting away. Lake Taghkanic was a much bigger and deeper lake back then. Its waters must have reached up to near the top of today's canyon. An enormous and erosive rush of water would have been flowing across a threshold here. Over a relatively short period of time, the erosion would have cut through the bedrock and created the canyon that we see today.

If you get a chance to visit the old canyon, take a few moments to imagine the surging white-water flow that was once here. This was another of those horizontal waterfalls, and it must have been a real

Figure 17-1. The Lake Taghkanic spillway.

natural wonder in its time. Today it is not given to us to fully enjoy this natural wonder; its time has passed, although we can see it and hear it in our mind's eye. But, as you can probably guess, there is a lot more.

We want to take you to a ravine just to the west of Lake Taghkanic and learn more of this area's marvelous history. That would be something called Doove Kill. We would like you to visit it, take a good look at it, and then, most importantly, *think about* what such a ravine really is.

To get there, find your way to County Rte. 15, just west of the park. Take it south to County Rte. 8 and follow that road west and downhill. Travel just about three miles and watch for Black Bridge Road. Turn right (north) there and you will very quickly encounter a bridge. The bridge passes over Doove Kill, and you can park near it, get out, and take a look. Here you can get your first glimpse of the ravine and actually contemplate what is here.

The features that most identify a ravine are the walls of rock that make up their steep slopes. You can see such cliffs here. Typically the creek is forced to pass across ledges of bedrock, and they break up its flow and produce white water, which helps the scenery considerably. Turn around and continue east on Rte. 8 for just about a half mile. You will soon enter the village of Snyderville, and there you will find Taghkanic Road. Turn left (north) onto Taghkanic and watch, again to your left, as you continue down that road; you will soon see more of Doove Kill (fig. 17-2).

For a short distance the road passes parallel and very close to the kill. Here it is a very real ravine; it meets all of the criteria. It is deep, steep-sloped, and has plenty of exposed bedrock. There are two small waterfalls here, but unfortunately they are on private property. Respect the private property and do not disturb the residents. There are, however, places where you can pull over, park, and gaze into the ravine without bothering anybody. It's a nice place, but watch out for the poison ivy!

For us to comprehend all of this, we have to go back in time about 14,000 years. That's when Lake Taghkanic was so much larger than it is today, and a lot deeper. The lake was swollen with the water melting off of retreating glaciers. Flowing out of that ice age version of the lake was an enormous volume of water. In short, Doove Kill had a very different personality back then. It was a powerful ice age torrent.

By now our exploring had shown us enough to rethink everything that we had just seen in today's park and along today's Doove Kill. In your mind's eye, we would like you to take those stretches of ravine on Black Bridge and Taghkanic Roads and fill them, almost to the top, with that one-time torrent. That makes Doove Kill something that you might call a Category Six white-water stream, or at least it was back at the end of the Ice Age.

Changes your impression of Doove Kill, doesn't it?

You might even say that, once again, this is something that rearranges your whole sense of reality. All of a sudden your image of the little kill has been dramatically altered. It has become a far more exciting place, one with a real ice age heritage. We have been seeing ice age spillways up in the Catskills. But Doove Kill speaks to us of a similar high-powered stream flow in the Hudson Valley itself. We are learning a lot about the past from looking at the present.

Figure 17-2. Doove Kill. Valley walls crowd the stream.

We all started out with an appreciation for the scenic beauty of something that we called the Taghkanic Park. But now we have learned something very different about this park. There is the notion that such a landscape has a geological heritage, and it can be a very rewarding experience to come to understand that heritage. This is, of course, the whole point of our book.

18

Bash Bish Falls

The Roeliff Jansen Kill is no Mississippi, but it is one of the largest rivers in the central Hudson Valley, and it does have an ice age story to tell. We thought we would be able to learn from the "Roe-Jan." We guessed that, if we just followed the old river and let it speak to us, then it would tell us its history. And it turned out we were right. The creek was very happy to speak to us, and we had a lot of fun listening.

One of the first things that people normally describe about a river is its source. The Roe-Jan has an inauspicious head in the southernmost part of the town of Hillsdale. It flows south from there and eventually becomes a real stream. But we found a better and more realistic beginning for this river when we looked at the map of its first major tributary. That's Bash Bish Brook, and we think it represents the real source of the Roe-Jan. Let's go there and take a look.

Bash Bish Brook originates in western Massachusetts and flows west. As it crosses the state border it flows through a very fine gorge (fig. 18-1); that's where the Taconic State Park is. The gorge is no accident; it is, we judge, a product of the Ice Age. This is where and when the

Figure 18-1. Bash Bish Gorge from above.

Figure 18-2. Bash Bish Falls from below.

Roe-Jan's story began. We would like to take you to the park as it was at the very end of the Ice Age, roughly about 14,000 years ago.

If you do go there, we would like you to picture the gorge as it was at that time. Up in the hills, behind the gorge in Massachusetts, there probably was still a lot of glacial ice, and it was melting—and melting very quickly. Vast quantities of water were pounding down the gorge. Bash Bish Falls is a pretty noisy place today, especially after a heavy rain (fig. 18-2). But back then, it was something else. You have to go there and let your imagination have free rein. Then, and only then, can you appreciate what is right in front of you. Do what we did; look up at the full expanse of the gorge and, in your mind's eye, fill it to the top with foaming white water. Make it loud, like a continuous explosion. Feel a pounding that would have almost made the ground shake. Bash Bish Falls is a scenic location; we are lucky to have it. But it has an ice age heritage that you have to know in order to truly understand it.

Let's keep going. Drive west to the village of Copake and then take Rte. 7A south a short distance, cross Bash Bish Brook, and look to your left and right. You will see a nondescript plain. But if you look carefully, you will notice that there is just the least bit of a slope dipping to the

Figure 18-3. The Copake outwash.

southwest (fig. 18-3). Geologists do notice such landscapes, and they speak to us. It is, we think, best described as glacial outwash; it's mostly sand and gravel that was washed out of the hills during the large rush of water that came at the end of the Ice Age.

Look at this landscape and see it again as it was 14,000 years ago. There is a rush of water coming out of the Bash Bish Gorge above. The water is brown, laden with sediment, mostly sand. The currents have broken up into dozens of small streams crisscrossing each other. We have seen these before; we call these braided streams (fig. 1-1). Braided streams are typical of situations where there is an overabundance of sediment, far more than the stream can carry. That sediment is deposited upon a barren-looking, glistening-wet, gently sloping plane, which is inclined in a downstream direction. There are few if any plants to be seen; they have not yet had time to colonize this fresh ecology. This is Bash Bish Brook as it was back then. From time to time there were even greater rushes of water out of the hills above. For brief periods of time, sizable sheets of water spread downstream across the whole surface. Each time, more sediment is deposited.

That's not the case any more. Long ago the glaciers melted and the braided stream that was Bash Bish Brook subsided to become the lesser flow of today. We are back in our own time.

19

The Northern Drowned Lands

In the last chapter we visited the upper reaches of the Roeliff Jansen Kill to learn about its ice age history. We traced the stream back to its origins above Bash Bish Falls and followed it to the village of Copake. We witnessed the melting of glaciers and the tremendous flow of meltwater that once rushed out of the Berkshires and into Columbia County. To see this is a privilege that comes with learning an area's geology.

But in this chapter we are going to see a very different sort of Roeliff Jansen Kill. If you look at a map of its drainage basin, from Copake to about four miles off to the west, you will find a landscape with many marshes and swamps. These we will refer to as the northern drowned lands. We have earlier described the drowned lands of Orange County, but the Copake wetlands are different. Here we will not encounter a sea of black muck, nor any substantial agriculture, but something very different. It is a landscape with a different ice age story. At the heart of this region is a parcel of land owned by the Columbia Land Conservancy. It is officially called the Drowned Lands Swamp Conservation Area. This is only part of the actual drowned lands, which cover much of Copake and most of northeastern Ancram.

The Roeliff Jansen Kill flows through the region. Here the stream's landscape is entirely different from anything seen upstream or downstream. The drowned lands are characterized by ponds and small lakes. The largest of these is Copake Lake, which you can see, northwest of Copake on Rte. 7. But we count at least a dozen others; most of them are well off the highway and out of sight.

The ponds and lakes are not the most important features in this stretch of the Roeliff Jansen Kill. Far more important are the numerous, and often very large, wetlands. Wander the roads of this area and you will commonly observe swamps, marshes, and bogs, big and small. All this we are hereon referring to as the northern drowned lands.

There is a hierarchy of terms that we use to describe types of wetlands. Swamps are just dry enough to support trees and shrubs without killing them. Marshes are so wet that trees and shrubs are excluded. Bogs are still wetter and, over time, they accumulate peat deposits. We expect that all three will be found frequently in this region. But the northern drowned lands are different from those of Orange County; here there is much more sediment and far less muck. These drowned lands display many swamps and marshes, but few peat-bearing bogs. At least, we haven't found them yet. The southern drowned lands are thriving agricultural lands, producing crops of onions and celery. The northern drowned lands have few farms; the soils here are apparently not especially rich.

So, how did these northern drowned lands come to be? What was their origin? To answer that, we have to go back in time once again to the end of the Ice Age. We have seen that vast quantities of meltwater were pouring down through Bash Bish Gorge and flowing out across the lands of Copake. Off to the west, starting in western Ancram, was a series of small hills. These impeded the westward flow of all this water, and much of it would likely have been pooled in the area of the northern drowned lands.

The New York State map of ice age deposits has this area mapped as outwash. We learned earlier that outwash accumulates south of moraines, and we did find at least one moraine to the north. We drove

Figure 19-1. The Drowned Lands Conservancy.

along County Rte. 7 and followed that moraine all the way to Copake. So we like the word outwash here; it does seem descriptive of what actually happened. Along many of the banks of the streams that flow through this area are exposures of fine-grained sand deposits. We have not spent much time studying these, but we are guessing that they are generally outwash deposits or shallow pond sediments and date back to post–ice age times. We are imagining the currents flowing out of Bash Bish Falls as overflowing the banks of their braided streams and spreading out across these drowned lands, carrying sand, silt, and clay with them. We like to use the word "puddling" to describe this. The Copake wetlands are remnants of this ice age history, but there is more.

As you drive through this area, try to imagine a few more feet of water covering all of the swampy locations. Go to the Drowned Lands Conservancy site and scan the horizon (fig. 19-1). See this vicinity as a fairly large, but very shallow, lake. If you want to, you can add a mastodon or two along the shores. That seems an appropriate image for the origin of the northern drowned lands of Columbia County.

It would take a lot of very strenuous fieldwork to properly document all of this. A geologist or soils scientist needs to hike about with a soil auger. He will stop here and there and drill holes into the ground to see the extent of the lake deposits. Over time, if he keeps at it, he can construct a map of the old lakes and ponds as they were. We wish we could do this, but we do not have the time and we certainly don't have the energy.

Still, we are fairly confident of what we are hypothesizing here. We travel the region of the drowned lands and we look into an ice age past. Then we see a landscape still struggling to overcome the effects of that history. We see Roeliff Jansen Kill drainage that has, even today, not yet developed enough efficiencies to get rid of all the water that has been trapped here since the disruption from the glaciers. That is a common theme for the whole region, and we will speak more about it in the next chapter.

20

The Wetlands

Glaciated landscapes, as we have gone to pains to emphasize, are wet. In our two chapters on the drowned lands, we have taken a look at the two most extreme examples found in the Hudson Valley. But there is much more. We would like to point out that there is a very common type of ecosystem that can be described by the umbrella term **wetland**, and it can be seen throughout the Hudson Valley. These wetland areas are commonly called swamps, and the definition of such ecologies is that they are very wet but not so wet to completely preclude the presence of trees.

People long regarded wetlands as synonymous with wastelands. Nobody could make any money from them, especially as they were usually not suited for any particular agriculture. History witnesses countless examples where wetlands were drained to make them "useful." The Drowned Lands of Orange County constitute a very good example of this. A substantial canal was dug there to drain off the land. Now those drained lands are agriculturally valuable. To a lesser extent, people have filled in wetlands to make them suitable for development and to stop the spread of mosquito-borne diseases. Those strategies were, especially during the nineteenth century, regarded as "improving" the land. Even the naturalist John Burroughs improved the land at his cabin "Slabsides" in order to make it suitable to the raising of celery crops. He was most enthusiastic about this sort of thing. That was in the late 1890s. Few notable naturalists would do such a thing nowadays, but attitudes were different back then. Burroughs thought he was being progressive.

Today we think that we have a more enlightened attitude about wetlands. We recognize that they do have an economic value, acting as areas of groundwater recharge and floodwater retention. These have ecological value, too. They are homes to countless plants and animals, including many rare species of reptiles and amphibians. They also provide refuge for birds migrating along the Hudson Valley corridor. In short, they are islands of biodiversity that today we believe should be valued and preserved. And they are abundant throughout the Hudson Valley.

We would like you to take notice of such ecologies. Few are open to the public, but they are there; however, most are unobtrusive. You can drive for miles, pass many swamps and take little, if any, notice of them. Still, we believe that once you have trained your eyes to take note of them, you will see them all over. And appreciate them.

One of our favorite wetlands is found in Columbia County, along County Rte. 27 west of West Copake. There, back during the late Ice Age, a heap of earth was deposited along a small tributary of Taghkanic Creek. That blocked flow from Chrysler Pond to the east, and with time allowed a good wetland to develop (fig. 20-1). The county highway runs right next to the wetland, so this is a convenient place to go and see one without getting your feet wet.

Now we have a better eye for wetlands, but what do they have to do with the Ice Age? The answer is, a lot. Surface waters of this sort are enormously abundant throughout all glaciated landscapes. After they have melted away, glaciers leave drainage in an awful state of disrepair. Heaps of earth left behind commonly block the pre-glacial flows of water. This damming of the landscape leaves abundant lakes, ponds, pools, swamps, bogs, and marshes. We noted that in both the Orange County and Columbia County drowned lands—moraines had blocked the old stream flows to create massive wetlands.

But in this chapter we are talking about something else, something smaller in scale but far more frequently seen. The term used is **deranged drainage**. A glaciated landscape is blanketed discontinuously with enormous amounts of nondescript heaps of glacial sediment. They do not have enough form to be called moraines, drumlins, or deltas. When they were deposited in contact with the melting ice, they can be called kames, but there is little shape to a kame, so it is not much of a feature.

But even if these masses of earth have little form to them, they do block valleys and other lowlands, and in so doing they derange

Figure 20-1. The Chrysler Pond Outlet swamp, located in Columbia County.

the drainage. Ponds, bogs, and swamps are created. Where there are streams, they are commonly very sluggish. Deranged drainage is common, and it is regionally important. The whole state of Minnesota displays deranged drainage. The state's license plates proudly proclaim the state to be the "land of a thousand lakes"—the state motto. It should actually be the "land of deranged drainage," but that might not attract as many tourists; it might instead scare them away.

And so it is that we might proudly proclaim the Hudson Valley as the "land of deranged drainage." But, again, this might not sound all that enchanting. Our point, however, still holds—there is an important ecology up and down our valley. It may not have been well regarded in past centuries, but we esteem it greatly today. And it is yet another of those many gifts of the Ice Age.

21

Niagara in Philmont

We have frequently worked with the Woodstock Land Conservancy, so we know what kinds of good work such civic groups can do. We weren't surprised, therefore, when the Columbia County chapter acquired a forty-seven-acre parcel of land, which they have opened up to the public in Philmont. The centerpiece of this site is High Falls (fig. 21-1).

The falls are well named; they tumble over a precipice about 150 feet high, along a river with the improbable name of Agawamuck Creek. We had not been aware of the site so, as you might guess, very soon the professors Titus were off to Philmont.

The conservation area is easy to locate off Roxbury Road, and there is ample parking. The trails are well marked (unless you are color-blind; they have red and green markers). We followed the green trail out to the overlook with its fine view of High Falls itself. We found it on a day when there was a strong flow of water, so it was a good view.

But we hadn't come to see the view; we were there to get a story. As we hiked along we passed several sizable outcroppings of bedrock. If you take one of the trails here, you will see fine-grained rocks with a dull sheen. These rocks are called phyllite. They are metamorphic; that means they had, during some ancient New England mountain-building event, been subject to intense heat and pressure. Metamorphism is, quite literally, the cooking of rocks. But, we had not come all the way out to

Figure 21-1. High Falls in Philmont. Photo by Steve Benson.

Philmont to do a story on dull-looking cooked bedrock; we needed something better than that. We would find it.

We hiked down on the upper blue trail to the bottom of the canyon. "Canyon" is the proper word for where we were. Very steep slopes, often with rock cliffs, towered above us. That's where we found our story.

The canyon took us back to the end of the Ice Age, about 14,000 years ago. That was, as we have been seeing all along, a very, very wet time in the history of the Hudson Valley. It would be fair to describe the whole landscape as soggy from the recently melting glaciers. Agawamuck Creek has a fairly low flow of water today, but back then it must have been a very powerful and erosive stream. That's when the canyon formed. That story is becoming repetitive, but here there is a new twist.

If you know anything at all about the geological history of Niagara Falls, then you will find the story of this little canyon to be a familiar one. Niagara Falls is a rather erosive site. Water flowing over the lip of the falls cascades downward and is very abrasive when it hits bottom. It carves out something called a plunge pool down there. The deep plunge pool undermines the stability of the cliff above, and eventually great masses of rock break loose and fall into it. Over long periods of time, and in this manner, Niagara Falls has eroded its way back, retreating upstream toward Lake Erie. Someday, Niagara will reach Lake Erie, at which point the whole of this great lake will drain into the St. Lawrence River system. That will be an exciting chapter in the history of the Great Lakes!

Well, our point is that High Falls has similarly been working its way back upstream. The canyon below Niagara is seven miles long; the one here is much shorter, only about a quarter mile in length. So High Falls is a scale model of the greater and far more famous Niagara, complete with a regulation plunge pool. We think that is notable, and that is our main story.

But there is a secondary story here; it's about hydropower. Philmont was once a mill town. The Agawamuck, back in the middle nineteenth century, was dammed and, along with some sluices, provided enough hydropower to support a number of mills. We drove up Summit Road to "factory hill," and there we saw a dusty old mill, built of brick, just above the falls. It is a well-preserved artifact of local nineteenth-century industry. It's been closed a long time. Perhaps that is unfortunate; we may be seeing global climate change associated with the burning of fossil fuels. Maybe the old mill should never have been closed. There is still a lot of hydropower here.

We can't finish this chapter without at least mentioning that the famed Kaaterskill Falls, across the Hudson Valley from Philmont, has also retreated, much as have Niagara and High Falls. It has managed about a half mile of retreat, probably over the same last 14,000 years.

22

Drumlins

In our rambles across the Hudson Valley, it has become apparent to us that **drumlins** are very important components of our local geology. Perhaps you have never heard of a drumlin, but you have passed by plenty of them. Drumlins are ice age features that are not usually very common. But nature has been generous around here, and she has blessed us with many of them. So, what are they?

Drumlin is a Gaelic word that means hill. But drumlins aren't just any hills. These are, for the most part, composed of coarse sediments, everything from clay to boulders. But there is much more. They have very specific shapes. We geologists like to say that they are shaped like inverted spoon bowls. They are the product of passing glaciers. The ice advances across the landscape and sculpts coarse glacial sediments into those spoon shapes. The upstream ends of most drumlins are usually relatively steep. As the ice advances, it typically molds the downstream end into a more tapered form with a lower angle. Left and right, the slopes are the steepest and most symmetrical. Take a look at our illustration and you will see a contour map (fig. 22-1) that depicts four drumlins of this shape. There is considerable variation from the usual expectations, but that is normal.

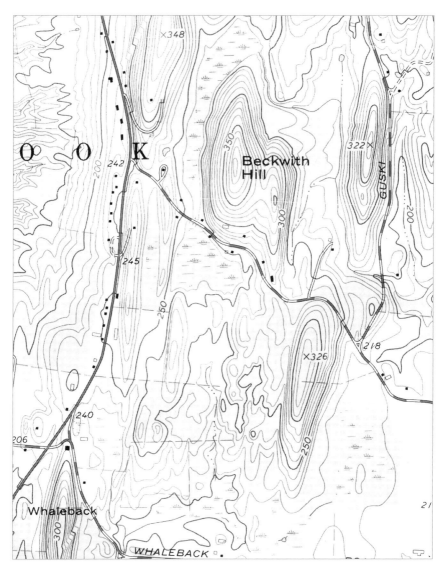

Figure 22-1. Drumlins. Map courtesy of U.S. Geological Survey.

Maps are good for finding drumlins, but it is better to go and see some real ones. We would like to take you to a good starter drumlin. That one is in the Columbia County village of Viewmonte, just north of Clermont. Take Rte. 9G about nine miles south of the Rip Van Winkle Bridge. Watch on the left (east) for Cemetery Road. Take that left and, less than a mile down the road, you will see a cemetery on the right. Be careful, this is easy to miss. Turn right into the cemetery and drive up that narrow driveway. You have climbed up onto a drumlin. You are driving up the north end of the hill, the steep side of the spoon bowl. When you reach the top, you will appreciate just how symmetrical a drumlin can be. Steep, but very smooth slopes form the two lateral sides of this drumlin (fig. 22-2). At the back of the cemetery, the driveway forms a turnaround, and there you have reached the tapered downstream end of the hill. This is a very good specimen of a drumlin; it has all the morphology that you would expect to see. Being a graveyard, the landscape has been kept cleared. There are relatively few trees and shrubs to block the view.

But drumlins are like potato chips; you can never have just one. If you explore the Hudson Valley region, you will find many more. In fact, we find clusters of them that can be called drumlin fields, and they are frequent. There is a good one just east of Valatie in Columbia County, and another just north of Poughkeepsie. Quite a few can be seen just east of the Hudson River in Germantown. Both Eight Mile Creek and

Figure 22-2. The crest of the Cemetery Road drumlin.

Basic Creek valleys at Greenville display serial drumlins. These Greenville drumlins are lined up at the bottom of the valleys and appear to represent the movement of currents of glaciers down those valleys.

That last observation is most interesting, but only hints at the formation of drumlins. It still leaves unanswered the question of how exactly they form. That has long been an issue debated late at night in geology bars. The big problem is that they form at the bottom of very large advancing glaciers. Unfortunately, it's hard to get to the bottom of an advancing glacier to see drumlins in the process of forming.

So, we are stuck—we have no firsthand observations to work with. We must make deductions based upon what little we can see. And what we can see is the pronounced streamlining of the drumlins. There is no escaping that they were formed by the sculpting effect of the advancing ice. The ice worked sort of like the hands of a potter on a pottery wheel.

But it must be more complex than that. There are some problems that this simple hypothesis does not explain. Here's the worst one: there is no question but that the ice advanced across Glacial Lake Albany. We have already explored the lake in this area, and the drumlins have been plopped down right on top of the lake beds, so they have to be younger, and maybe a lot younger, than the lake deposits. But, if that is the case, then where did the drumlin sediment come from? Lake beds are fine-grained clays and silts, so they could not have been sculpted into drumlins. The question is a tough one with no obvious answer. That happens in science.

Even if there are mysteries that remain, we are confident that, once you have learned to spot drumlins, you will enjoy seeing them and will have that chance very frequently.

23

March of the Glaciers

We have become increasingly surprised at how many drumlins there are hereabouts. They are striking features when viewed from the ground or air and, once you have learned to recognize them, you too will be surprised at how many there are. New York State is renowned for this within geological circles. The fact is that the lion's share of the *world's* drumlins is found in our state. Makes you swell with regional pride, doesn't it? Well, actually, it should.

Seeing them on the ground is one thing, but here we want to work from a graphic (fig. 23-1). We present a satellite image of Greene County in the upper Hudson Valley, and it begins to portray just how recognizable and frequent drumlins are. Take a good look; see all those spoon-bowl-shaped features down there? Each is a drumlin. Notice that they are oriented parallel to each other. That reveals the direction of the flow of the ice that made them. In this case the right arrow portrays a glacier traveling down the Basic Creek Valley. This was likely very late in the Ice Age, and this particular glacier was probably not very big, nor very thick. It could easily be steered by the valley. To the

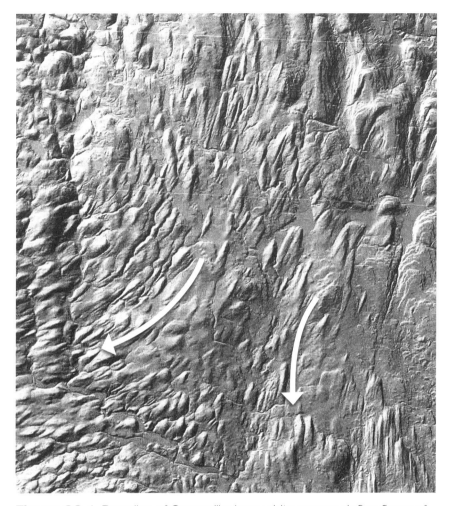

Figure 23-1. Drumlins of Greenville; long white arrows define flows of ice in valleys. Courtesy of U.S. Geological Survey.

left (west), the second arrow shows a complex of drumlins that was produced by a second glacier, one that descended the length of the Eight Mile Creek Valley.

Our second graphic (fig. 23-2) is an outline map of Columbia County. It's adapted from the New York Geological Survey topographic maps of our region, and it shows just how frequent and important drumlins are in our region. Each dash is another drumlin. We have also added a map-

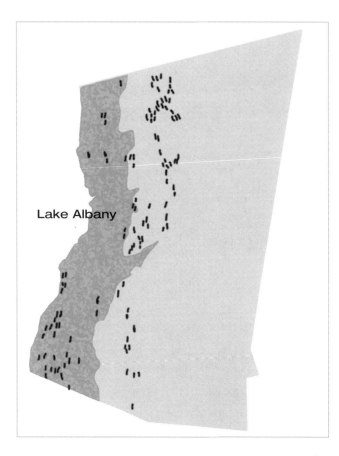

Figure 23-2. Drumlins of Columbia County.

ping of Lake Albany here as well. We can learn to appreciate what these maps represent and the history they display.

Our map shows a host of drumlins. We count well over 100 of them. Each dash is parallel to virtually all of the others. Each one has a long axis that records the southward direction of the glacier's flow. Collectively they define the passage of a glacier that, though big, wasn't as large as some of the others we have seen in this account. It was just small enough to be channeled by this part of the Hudson Valley. You can "see" its flow from the dashes. While most of the motion is due south, some of the ice veered to the southeast.

This flow of ice is not nearly as large as the Hudson Valley glacier that we saw early in this history. We are uncertain of its full extent. The many drumlins you see on the map cover all of Glacial Lake Albany and stretch a little to the east of the lake's old eastern lakeshore. Evidently, the ice had little difficulty moving across the old lake's deposits. The landscape here was smooth and flat, and offered little to block the flow. To the east the landscape is a little more elevated, and the ice only pressed a short distance that way. Thus the ice was essentially confined to the central Hudson Valley. The glacier evidently thinned to the south. Around Kinderhook drumlins reach elevations up to 650 feet. That drops to between 300 feet or so from around Clermont to the south.

You can understand why geologists get excited about this sort of thing. With maps like these we can closely document the last stages in the great glaciation that brought the Ice Age to an end in our region. Our drumlins bring the glaciers to life and graphically speak to us of what the glaciers were doing.

And our story has a lot of detail to it. The ice had previously retreated to somewhere in the north. This had left behind Glacial Lake Albany, which filled most of the Hudson Valley. But the drumlins are features that lie on top of the lake beds; they must be younger. That means that there had been one last readvance. When, exactly, this happened we cannot say with precision. The advance came after the time of the lake, but how many years or decades or centuries later we do not know—probably not all that many.

We are forced to depend on our imaginations a little here, at least until better evidence turns up. It is even possible that the lake was still there as the ice advanced. It would be easy for the ice to "skate" across the frozen lake. That would account for its apparent confinement to the lake area. In science this sort of imagining is called "hypothesizing."

Sounds a lot better than "imagining" or "guessing," doesn't it?

Remember the last time you skated upon a large pond, or better, on a large lake? Now, in your mind's eye, look north and watch as a glacier moves slowly toward you over the lake. It's quite a vision, and the evidence tells us that such a thing just might have actually happened.

24

Ice Age Architecture

The Titus family estate consists of fourteen acres of mostly forest, surrounded by more forest. The house abuts a rural highway and overlooks Catskill Creek. Beyond is a view of the Catskills. We have been blessed with the good fortune to own such a nice property, and we have the good sense to know it.

There are, however, responsibilities that come with the stewardship of such land. Where, exactly, should the nature/ski trail be laid out? Where should we open up views of the mountains? *Should* we do so? Is it practical to make a little frog pond? Should the forest be thinned out or left to its own? In short, land ownership is work!

But it can be an art as well. Our little forest is something of a Hudson River School of art painting in progress. Our woods take us through an Asher Brown Durand sort of forest, and we value that. Sunset over Windham High Peak is a Frederic Church moment if there ever was one.

We have been learning firsthand something about landscape architecture. And that's a field that has a venerable history here in the Hudson Valley. Many of the great nineteenth-century pioneers of this field practiced here. And you can go see much of their work at the many great estates that are open to the public.

There is probably no better place to start than at Olana, Frederic Church's Moorish Revival mansion and estate south of the city of Hudson. Church was the greatest of the Hudson River School painters, especially from the 1850s to the 1870s. Tragically, however, his talented hands were struck with rheumatoid arthritis, and his career too quickly entered into a long decline.

But, if Church could not paint, he could still go on being an artist, and his 200-plus acres, located atop Church Hill, would be the canvas of his second career. Church spent the last quarter-century of his life preening his estate. It became an artistic kingdom of planned views and designed scenic settings. Sadly, much of Church's work has come to be overgrown. Still, a lot of his landscape architecture remains and, if you have not done so already, you should go and explore Olana's many trails.

Church had a fascination with the explorations of the Arctic that were underway in his lifetime. He likely knew that glaciers had once overrun the Hudson Valley, but he probably would have been surprised had he known just how much of his landscape was glacial in origin.

Think about it: glaciers, three or four thousand feet thick, slowly advanced southward, scraping the ground beneath, sculpting the whole Hudson Valley. Church Hill, and nearby Mt. Merino, especially felt the results. They became what geologists call "**rock drumlins**." That means that they are bedrock hills that became streamlined by the passing ice (fig. 24-1). That's unusual; as we have seen, most drumlins are made of coarse sand and gravel. All of them, including those made of rock, have the form of an inverted spoon bowl. All are streamlined so that the down-ice end is stretched out into a very gradual slope. The northern, usually upstream, end is steeper. They generally display beautiful, sinuous curvatures, and they can make very handsome landscape features.

We have seen hundreds of sand-and-gravel drumlins in the Hudson Valley, but Church Hill and nearby Mt. Merino are the only rock drumlins that we know. Olana sits at the very top of one of them. The estate has a good view of the other. Church's primary planned view is the one from his

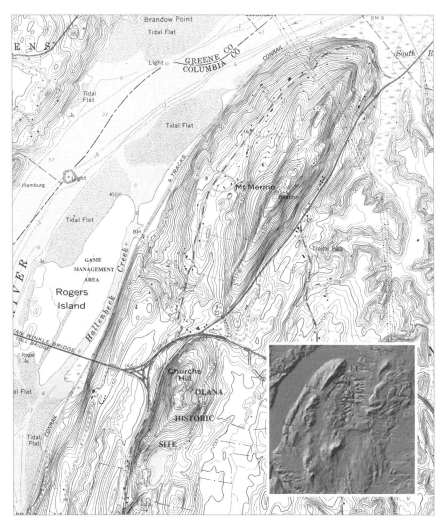

Figure 24-1. The rock drumlins of Church Hill, south, and Mt. Merino, north. Insert: satellite image of Mt. Merino and Church Hill. Courtesy of U.S. Geological Survey.

---

south-facing porch. It is a truly breathtaking, sweeping panoramic view of the Hudson Valley and Catskills beyond (fig. 24-2). It was designed to be the centerpiece of the Olana property, and it is. It draws your eyes down the drumlin's gentle southern slope and then off into the distance. The view continuously changes as the sun makes its daily migrations

Figure 24-2. The view from Olana.

across the sky; it varies with the seasons and with the weather; it is never the same two days in a row, or two hours in a row. It is a wonder; it is a gift of the Ice Age.

In the end, none of these aesthetics are important to the *science* of the Ice Age. Landscape architecture is an art, albeit informed and guided by science. Its efforts enrich our understanding of the ice age origins of such a historic site by focusing our attentions upon selected details of the landscape.

But, up and down the Hudson Valley, there is so much more.

25

Valley of the Kings

The east bank of the Hudson has long been the fashionable side of the river. It was once, and still today remains, the home to some of America's oldest and most respected families. The culture of that society expanded during the nineteenth and twentieth centuries and may be best remembered from the works of Edith Wharton, who modeled her characters after those real-life Hudson Valley denizens in books such as *The House of Mirth* and *Age of Innocence.* And what an aristocracy it was! From the Van Rensselaers, the Livingstons and the Montgomerys, to the Roosevelts, Astors and Vanderbilts, all built grand mansions along the east bank of the Hudson. The region rivaled Newport, Rhode Island.

Most of these mansions still stand. Many are well preserved and open to the public today. As residents of the region, we have long enjoyed visiting those old "American Palaces": Clermont, the Mills Mansion, Montgomery Place, Springwood, Hyde Park (the old Vanderbilt home), Wilderstein, Locust Grove, and numerous others. But why the *east* bank of the Hudson River? Was it just a fashionable place, or is there more? On our travels we have stumbled upon a possible answer—discovered, of course, in the area's ice age geology.

You have to be able to look at the Hudson River, not as it is today, but as it was in the distant past. You must do what we have been doing throughout this book and go back about 14,000 years or so. That's to a time near the end of the Ice Age. It was quite a scene; much of the lower Hudson Valley had recently been deglaciated. To the north, however, a great ice sheet still extended from the Albany area all the way to the North Pole and beyond.

But, as we have seen in chapter ten, the valley was beginning to recover from the Ice Age—not so much from the cold as from the weight of the glaciers. You must remember that thousands of feet of ice had flooded the Hudson. The weight of this was enormous, so great in fact that, as we saw earlier, it compressed the very crust of the Earth itself, pressed it down a hundred feet or so. It would take a great deal of time for the crust to readjust; we called it "isostatic rebound." By about 16,000 years ago the southern Hudson Valley had already managed much of its rebound, but the middle Hudson was still depressed.

This depressed zone formed a basin for Lake Albany, and quite a great thickness of lake bottom sediment accumulated along with more than a few deltas, especially along the lake's eastern shore. Given time, the middle Hudson Valley completed its rebound, rising substantially. All of the waters of Glacial Lake Albany were dumped into the Atlantic, and the river developed its modern flow. As it happens, the river chose a course mostly on the west side of the Hudson Valley, leaving behind many of the old lake deposits to its east. As it resumed a normal flow, the river's currents cut into the old lake deposits. This left a peculiar east bank landscape. Well above the river there came to be a fine, flat, elevated plain—the old lake bottom. The Hudson cut into this, often leaving steep, river-eroded slopes. These would display wonderful views of the whole valley and attract numerous home builders.

Early on, when the Livingstons came along, they parceled out many of the choicest sites for their homes. Later, in the nineteenth century when the industrial age millionaires followed, shopping for good real estate, this setting was ideal. The resulting mansions were placed on the remaining high, flat bluffs. They were thus provided with fine vistas of the river and the whole of the Hudson Valley beyond. The sites were visually stunning. Not surprisingly, the area gradually filled up with stately homes and mansions, and that distinctive Hudson Valley aristocracy continued to evolve here.

And they hired the best architects, including pioneers in the field of landscape architecture. Men such as Andrew Jackson Downing, Alexander Jackson Davis, Frederick Law Olmsted and Calvert Vaux practiced here. Samuel F.B. Morse, mostly remembered as the inventor of the telegraph, was also an accomplished artist and amateur landscape architect.

See for yourself. We have already visited the Vanderbilt Mansion and the Roosevelt home situated on the Hyde Park Delta. But there are other fine homes, open to the public. Let's highlight two of them.

First, travel to Annandale-on-Hudson, just south of Bard College, and visit Montgomery Place. Explore this large, complex, and picturesque estate. As you approach the mansion from its driveway, make note of the flat landscape upon which the home is situated (fig. 25-1). This surface is the bottom of Lake Albany, elevated to today's perch by that isostatic rebound.

Typically for Hudson Valley picturesque architecture, the grounds near the house are very well manicured. But, farther away, the land was allowed to be more natural. That's quite intentional and, again, typical of the architectural style of that era. Toward the river the lawns slope away from the house, and stone walls interrupt the slopes, forming terraces. Off in the distance, where the river has cut into the lake deposits, there are steep, forested slopes. Again, this is mostly intentional. The plan was to surround Montgomery Place with very well-manicured landscape that would grade slowly into a natural or picturesque more-distant border.

There are two places on the border where trees have been intentionally cut, and these spaces provide the planned views. The first was opened up by cutting the forest northwest of the house. That afforded a splendid vista of the northern expanse of the river, with mountains behind and the Sawkill Creek in the foreground. The second

Figure 25-1. Montgomery Place sits upon the floor of Lake Albany.

planned view faces due west. It displays the river in the foreground with a series of hills in the far distance. If you visit here, please wander the grounds and explore its many trails. But, most importantly, look up and down the river and enjoy the views. It may have been our unique Hudson Valley gentry that built the house here, but the glaciers did most of the rest.

South of Rhinebeck is another fine mansion, called Wilderstein. There we see a much younger house, but with very much the same landscape architectural strategies. As at Montgomery Place, the most-cared-for grounds are all right around the home itself. Away from this centerpiece, the landscape gradually becomes more picturesque and natural. Wilderstein, too, is situated upon a platform of Glacial Lake Albany strata and, again, those deposits are intersected by steeply eroded riverside slopes. Most are forested, but one stretch has been cleared to provide a splendid outlook facing south and overlooking the Hudson (fig. 25-2).

As is commonly known, the lower half of the Hudson is technically not so much a river as it is an estuary of the sea. The Hudson, throughout most of its flow, has been forced to cut into its tall banks of ice age deposits. And we have found that this apparently accounts for much of the Hudson's special scenic nature. This river is thus very different from most, and perhaps all, other rivers. It may be unique. Therefore

Figure 25-2. The view from Wilderstein. The platform the house is perched upon is also the bottom of Lake Albany.

it should not be surprising that a distinctive landscape architecture should have developed here. Much of what we in America know as landscape architecture has descended from our valley, and this may be yet another of those many gifts of the Ice Age.

26

The Post-Glacial Fractures

We could use some help. We have spent a lot of time researching this book, but we haven't been able to solve all of our problems or even find all of the features that have been reported. There are some very interesting post–ice age features reported in the literature that we simply have not been able to find. Maybe, if we describe them, you can find them for us. Our features are called **post–ice age faults**. These are fractures of the bedrock that occurred during the isostatic rebound of our Hudson Valley crust. You will remember from earlier discussion (chapter ten) that the advancing ice caused a depression of the whole landscape. After the ice melted away and the weight of it was removed, the crust rebounded. The amount of rebound was quite substantial, and that must have generated a considerable amount of stress within the bedrock that was affected—and that would be pretty much all of the bedrock of the Hudson Valley.

Such bedrock is brittle. If it is pressed down and then rebounded, chances are that it will have been subject to so much stress that it would have fractured. This certainly happened, commonly and repeatedly,

Figure 26-1. Early-twentieth-century photo of fractured bedrock. Courtesy of New York State Museum.

---

throughout the Hudson Valley, but the question is how do we recognize *post-glacial* fractures? How are they different from any other cracks? Rocks are very frequently fractured; how do we spot such young cracks? The answer is that we watch for bedrock surfaces that have been planed off, scoured and striated by the advancing glaciers (chapter three). If such rock happens to also be fractured, and those fractures pass across the striations, then the fractures must be younger than the ice age features that are broken by them. They thus must be post–ice age features.

These, it would seem, are common. We have found references in the literature by writers who had no trouble finding them. But we have gone looking and come up empty-handed. So, we are asking for your help. The literature tells us that these post-rebound fractures were first discovered by William Emmons in 1841. He found them east of Hyde Park. They were reported to be common in the vicinities of Copake and Ancram. J. B. Woodworth of the New York State Museum took an interest in them during the early twentieth century, and he found them in those vicinities. His colleague, a Professor Merrick, found them in the town of Clinton. Woodworth did a lot of exploring and found more such faults in Rensselaer and Troy, too. Good photos are available showing them breaking the surfaces of outcrops lying alongside of the roads in these towns (figs. 26-1 & 26-2). But these discoveries occurred in the early twentieth century, and those roads have been widened since then. The fractured surfaces would have been obliterated by the roadwork. Or at least that's what we think. Maybe we are just not very good at finding such things.

The types of rocks that preserve such fractures consist of the relatively soft shales, slates, and dark sandstones that are so common in the Hudson Valley just east of the river. These rocks were easily beveled and

Figure 26-2. Post–ice age faults along the road on the right. Courtesy of the New York State Museum.

scoured by the passing ice. Then, too, these rocks were particularly brittle at the time of the rebounding; fracturing was easy for them.

So, we are asking you to be on watch for such features. Exposures of glacially scoured bedrock are commonplace on ledges that lie alongside of highways. See a good one? Stop and take a look. We could use the help!

27

Silent Earthquakes on the Hudson

The heavy rains in the autumn of 2003 picked up where the Ice Age had quit and began a slow earth movement that would doom a whole neighborhood on 1st Avenue in Schenectady. But it was not until the next spring's thaws that anybody knew about the problem. One morning people woke up and found that a sizable crack was opening up in the land behind their homes. Over a period of several days the crack widened. Gradually the ground began to sag beneath six 1st Avenue houses. This was not your standard Hollywood movie catastrophe. There was no sudden wrenching of the landscape. There was little if any noise. It is likely that, at first, many people were not even aware of what was happening beneath their feet. But it was a catastrophe nonetheless; soon about a dozen families who lived there had to be ordered out of their homes. They became refugees from what are called "geohazards." Their neighborhood had been destroyed, just as if hit by a volcano or earthquake.

We have, over the years, been watching the news of multiple events such as happened in Schenectady, and we have had a growing sense that something has been going on geologically here in our middle and upper Hudson Valley. We think a pattern has been developing. Scientists notice patterns, and they seek to understand them. This one *needs* to be noticed and explained.

This is not just a matter of coincidence or bad luck, something that just happens from time to time. That is not the case here; we fear that the Schenectady landslide is something that can happen in many neighborhoods up and down the Hudson Valley. And few people are aware of the danger.

The Hudson Valley is a remarkably stable location geologically. Real earthquakes are very rare and, when they do strike, the force released is minimal; the damage is hardly to be noticed. There haven't been any volcanoes in this area for about a quarter of a billion years. But the earth movement of Schenectady is something different. This is something that actually does happen here. It is sometimes called a "silent earthquake."

A silent earthquake occurs when there is a heavy, waterlogged mass of earth adjacent to a steep bank. The weight builds up stresses within the ground and creates a concave-up, curved fracture (fig. 27-1). The fracture might remain stable for long periods of time but, eventually, unusually heavy rains come and the water soaks into the ground, concentrating along the surface of the fracture. The pore pressures build up

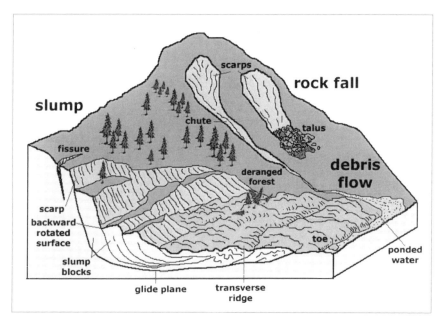

Figure 27-1. Slump diagram. Courtesy of the U.S. Geological Survey.

within the fracture, and that surface becomes slippery. Then the ground above it begins to sag. The earth slides as a great, single, cohesive mass, rotating along the curved fracture. Trees lean backward. Geologists call this type of movement a "slump." Think of it as a slow-motion landslide. It is more technically called a **rotational slump**.

The reason that the whole Hudson Valley faces prospects of such slumping goes back to the Ice Age. As we have seen, between 13,000 and 15,000 years ago Glacial Lake Albany flooded much of the valley. Where rivers flowed into that lake, they created deltas. The lake drained long ago, and now the old deltas have been left high and dry (chapter eleven); we call them "hanging deltas." A hanging delta is a heap of sand, silt, and clay. It is elevated above the surrounding landscape, and therefore it is typically safe from any flood threats. The earth is well drained most of the time and, all in all, the delta makes a fine place to build houses—or so it would seem. Not surprisingly many towns and cities are located on old ice age deltas. They just appeared to be good places for development, or at least they did one or two centuries ago when development began. Schenectady is one of these "favored" locations, and so too is Rotterdam. They are built upon one of the largest of those deltas, the one constructed where the Mohawk River emptied into Lake Albany (fig. 27-2).

The problem with deltas is that they commonly have steep foreset slopes all along their outer edges (figs. 11-1 & 11-5). The Schenectady slump occurred at the very edge of the Schenectady Delta. That delta is about 100 feet tall, and 1st Avenue lies nearly 100 feet above the bottom of the slope. The 2004 slump was inevitable as heavy rains soaked into the ground and destabilized the outer edge of the delta so that gravity could eventually pull the earth, and the neighborhood, downhill.

As we have seen, the list of cities and towns built on old ice age deltas is a long one. In addition to Schenectady and Rotterdam, much of Saratoga Springs is on a hanging delta. There are several more to the east at Schaghticoke, Schuylerville, and Kinderhook. To the south there are more at Catskill, Rhinebeck, Hyde Park, and Poughkeepsie (fig. 11-7).

Well, you see the problem. All these deltas have flat tops. They seem like such nice places. Delta tops never see floods; it is easy to dig base-

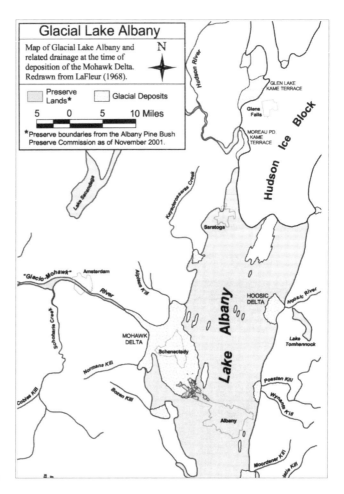

Figure 27-2. The northern part of Lake Albany, late in the Ice Age. See the Schenectady Delta. Courtesy of New York State Museum.

ments into them. It is also easy to lay out road grids. All of these apparent benefits encouraged eighteenth- and nineteenth-century development, which continued on into the twentieth. Few people, if any, understood the ice age history here. It is not likely that anybody sounded an alarm. That's unfortunate.

The worst locations are found where there are steep delta foreset slopes. First Avenue in Schenectady was a classic. This steep slope was a feature of the ice age delta itself. When the rains came, that slope was primed and ready.

28

Yellow Alert?

As we said earlier, we have been watching a number of these damaging slumps in the upper Hudson Valley. These have been front-page stories in the local papers. The first one we took real note of was in 2000—the Delmar slump, south of Albany, which put a major road out of commission for quite some time. It had been built on the muddy sediments of Glacial Lake Albany. The sediments simply gave way and slid into the Normans Kill. Well, these things happen, or so we thought at the time.

We watched as another slide occurred in Greenport, across the river, just east of Hudson. Soon we had a small slide just a mile from our home in Freehold. Again, in the spring of 2004, we saw still another nearby bank give way, and since then it has been oozing water. That's too close for comfort.

Slumps are an ongoing problem in the Hudson Valley, but there seem to be a lot of them in recent years. We took note of them. We saw another one occur in Amsterdam, then still another in Rensselaer. These also seem—all of them—to have involved the sediments of ice age lake deposits. That was alarming to us; why were these events coming at such a rapid rate?

But then it got even worse. We began hearing from people all over in response to all the newspaper geology columns that come out of our home. We began receiving e-mails from people in Valatie complaining about flooding basements. Three houses on New Street there had been experiencing serious problems for weeks. Basements flood—that's their job, but some of these folks claimed that they had never seen the likes of this even after decades of residence, and they were frankly alarmed. We heard much the same from people in Kinderhook and several other towns.

All this may just have been coincidence and might have meant next to nothing. Or, all this may just have indicated that we have had a lot of rain in recent years. That would explain one or two year's problems, but it would not tie in all the events of recent times.

In the end, it seemed to us that there was enough to warrant a little investigation. It looked on the face of it as though the region's water tables have been rising and the recent heavy rains have triggered a series of landslide problems. This trend may be something that has been developing over the last several decades. Could we document this the way scientists should, and could that lead to an explanation? Well, we could try.

We checked with the National Oceanic and Atmospheric Administration website and found some interesting things. New Yorkers have seen some climate change over the past century. Our average temperature has climbed, but only about one degree Fahrenheit. More interestingly, however, our rainfall has climbed about six inches, from 36 to 42 inches/year. That's a 16 percent increase, and that is a lot. The year 2011 saw rainfall nearly 40 percent above average!

If we have seen a lot more rainfall, then it follows that there should be more groundwater and higher water tables. Make things worse with a few heavy rains, and it seems logical that basements would start to flood and slumps might be triggered. People might well remember that these things didn't happen in the distant past because they really couldn't have.

What we are suggesting is that if we have a wet summer—or, worse, a snowy winter and rainy spring—any time in the near future, then we may see serious problems. Is all this good science? Certainly not; it is the result of just a little work over a short period of time in response to some rapidly occurring events. It's not scientific theory, just hypothesis, but it needs to be taken very seriously in our Hudson Valley.

What is the nature of the climate change that we may be witnessing? One possibility is something called the **North Atlantic Oscillation** (NAO). This is a major weather pattern that is associated with the distribution of regional air pressure systems in the North Atlantic. Sometimes a great high-pressure system sets up atop the Azores while a sizable low-pressure system lies nearby to Iceland. When that is the case, then there is a rotational motion to the North Atlantic weather. The masses of air migrate in a clockwise fashion. This brings unusually large nor'easters up the Mid-Atlantic coast. It rains a great deal in coastal New England. Rainfall is especially heavy in March and April,

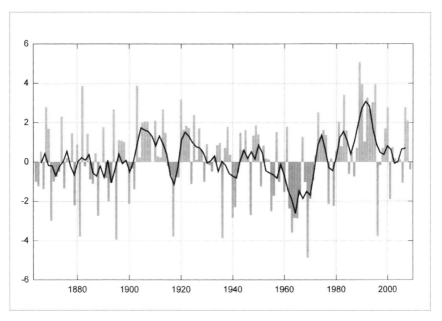

Figure 28-1. Graph showing ups and downs of the North Atlantic Oscillation during the last 150 years. It essentially indicates that dry periods occurred from 1950 to 1970, with wet climates until 2000 and the possibility of a return to dry periods in recent times. Courtesy of National Center for Atmospheric Research.

cool seasons when a low rate of evaporation can result in heavy waterlogging of the soils. That's the time of the year when earth movements such as slumps most frequently occur. Most of the damaging ones that we have witnessed occurred during the early spring season.

Is the North Atlantic Oscillation related to global warming? That issue is debated, and we voice no opinion. For us what is important is that over the course of the past half century there has been an increase in rainfall of about 16 percent in the Northeast, and that is troubling. It certainly goes a long way toward accounting for all the slumps and wet basements that have been observed.

What we are hypothesizing is that a broad climatic pattern has set in. This pattern, the NAO, has triggered instabilities in ancient ice age deposits, and those have led to landslide events. In that light it is curious that, just as we have been preparing this book, the effects of the NAO seem to have started abating. Currently the cycle seems to be reverting to a condition that will result in fewer winter coastal storms and, perhaps, a return to drier times. If this is so then ... never mind!

29

The Slides of Hyde Park

Some of the Hudson Valley's most historic locations face, it would seem, the potential threats of slumping. We have seen that the town of Hyde Park is built upon the delta of the stream with that strange name of Crum Elbow Creek. Most of the town is a broad, flat platform. That platform marks the approximate level of Glacial Lake Albany when it was flooding this part of the valley. Crum Elbow Creek deposited the sandy sediments of its delta right up to the water level. As the delta grew, it pushed out into the lake but maintained a level that today (after isostatic rebound) lies at 180 ft. After the Ice Age was over, it was only a matter of time before the waters of Lake Albany would drain. They did, and Crum Elbow Creek proceeded to cut a canyon right through the heart of its own delta and what is today Hyde Park (fig. 11-4). The rest of the flat delta top remains in good shape (fig. 11-3).

The front of this delta, as with any typical delta, is a very steep slope, and much of it overlooks the Hudson River. This directly led to the siting of these historic sites. As we have seen, when the great wealthy families of the Gilded Age settled along the Hudson, they sought out properties

Figure 29-1. Springwood, at the edge of the topset of the Hyde Park Delta.

with great views. The Vanderbilts built their mansion "Hyde Park" at the crest of the Crum Elbow Delta (fig. 11-3). It has a wonderful sweeping view up and down the Hudson (fig. 11-2). Similarly, the Roosevelts built their home "Springwood" to the south, also on the crest of the delta foreset (fig. 29-1).

Those two mansions now are open to the public and are enjoyed by thousands of visitors every year. Springwood is the cornerstone of the Roosevelt Presidential Library. These are, in short, very important places, but they may both someday tumble into the Hudson—or, at least, they may tumble *toward* the Hudson. Like any steep slope, the crest of the Crum Elbow Delta is likely to be landslide-prone. As in Schenectady, curved fracture surfaces have apparently in the past developed along the edge of the delta. They have been active at times, and periodically slumps have occurred all along the delta front including the two mansion sites. You can go and see this for yourself.

Figure 29-2. "Scallops" south and below Vanderbilt Mansion.

The Vanderbilt home is an enormous property with a fine system of trails and driveways. You can hike downhill at the south end of the mansion and work your way along the bottom of the delta. If you do this, you will see that the slope here has a scalloped appearance (fig. 29-2). Each scallop looks as if a giant ice cream scoop had taken a mass of earth out of the ground. There are more scallops just north of the mansion. The lower slopes here, below many of the scallops, have a wrinkled look to them (fig. 29-3). These wrinkles appear to be heaps of earth deposited where the ancient slumps ended up.

The upper reach of these slumps is called its "head." Each has a semicircular shape, and those compose the scallops we see here. Each wrinkled lower section is referred to as the "foot" of a slump. Once you become familiar with this structure, you will soon appreciate that they are commonplace along the western edge of the delta. If you walk south from the Vanderbilt Mansion toward the formal gardens, you will find one

Figure 29-3. "Wrinkled" slump structures, north and below Vanderbilt Mansion.

slump structure after another. Here the steep delta front is forested, but you can still see its structure.

Likewise, if you visit Springwood you can head south along the crest of the delta and see a scalloped slope extending parallel to the river. The landscapes of both homes suggest an extensive history of slumping. And, ominously, some of those slumps reach right up to the mansions themselves. The situation at Springwood is the most frightening. Steep slump slopes have the house surrounded on three slides.

Both the Vanderbilts and the Roosevelts chose to position their homes as close as was practical to the edges of the slopes. They no doubt wanted to maximize the picturesque views of the Hudson Valley. The architects of the time apparently did not understand the ice age history of these properties, and they certainly did not recognize the scars left by a history of slumping. Both mansions have stood for a very long time, but we fear that both will not last; each will someday begin to tumble toward the river. These homes were, in fact, built as close as *impractical* to the slopes.

At least, that is our hypothesis. There is a way of testing it and evaluating how likely such slumping is. As we understand it, an active slump will display certain patterns in its trees. Typically, the trees at the top of the slumping will be inclined backward. Slump has "pulled the rug out from under them," and they are falling backward. At the bottom of an active slump, delta materials are thrusting forward and downward. Thus, trees at the bottom of a slump will be shoved forward and leaning downhill. So, at the top of the slump, trees are leaning backward, and at the bottom they should be leaning forward. That provides us with a way of testing our hypothesis.

We did a lot of hiking and kept looking at the trees. As expected, some at the top did lean backward (fig. 29-4), and some of those at the foot of the slope did lean forward. But there simply were not all that many leaning trees. The slope would thus *appear* to be relatively stable. Perhaps the danger is not all that great. We are cautiously hopeful that no terrible slides will carry away either of these fine old mansions, in the near future. It might be advisable, however, if people saw to it that there

Figure 29-4. Trees tilt backward from the sliding, near Vanderbilt Mansion.

was good drainage. We did see evidence of drainage pipes being installed at the Vanderbilt Mansion, and that is reassuring—we hope.

    This is serious business. We have seen slumps along other deltas; we cannot dismiss the danger here. And perhaps the most frightening thing is that there really is no way of guessing when the slumping might return. The silent earthquakes of the Hudson Valley do not provide much in the way of warnings. They do come after very rainy seasons, but they strike first here, and then there, without much logic. By the time we know there is a danger, the worst may well be over.

30

Elephant's Graveyard

In recent years there has been a small flurry of discoveries of ice age elephants in New York State. You may have read about the mastodon discovered in Hyde Park; it made quite a stir down there. The bones were found quite by accident in a swampy section of a family's backyard. All they had wanted was to dig a small pond; what they got was an ice age treasure. Researchers from the Paleontological Research Association in Ithaca spent the summer excavating the skeleton, and most of it was recovered. Such events are very exciting, and people came from all over to see the bones as they emerged from the mud. Some even got involved in the project. There is never a shortage of local volunteers to help out on a dig like this.

Less well known was a similar discovery near Ithaca. Another swampy area yielded the bones of another ice age elephant. This one was a woolly mammoth, and this dig attracted dozens of Cornell students. A big surprise awaited them when, during the excavation, the remains of a second skeleton, this one being of a mastodon, came to light. You can imagine the excitement that accompanied this "twofer."

Such discoveries are always big news stories in the local area. They should be; they are rare and exciting events. But, during the eighteenth and nineteenth centuries, a great many more such discoveries were made (fig. 30-1). In those times it was common for farmers to drain their swamplands, and this frequently led to the discovery of large bones. The Hudson Valley of New York State became something of a world capital for mastodons, as these distant relatives of the modern elephant were apparently very common here, especially in the drowned lands of Orange County in the lower Hudson Valley. The hunting has not been nearly as good in the upper Hudson Valley, but some very historic finds have been made there.

What may surprise you, as it did us, is that it was in the Hudson Valley, specifically Columbia County, that a mastodon skeleton was first found. This story takes us all the way back to the year 1705. That's when a mastodon tooth was found by a Dutch tenant farmer in what was then called Claverack Manor, a site that may well be located in Greenport today. Spring floods washed it out of a bluff about sixty feet above the Hudson River. The remains were fragmentary, but they impressed the English colonists of the time. The big find was a tooth that weighed almost five pounds. Today such a tooth would be quickly identified as belonging to a mastodon, but back then nobody knew of such animals. There was great debate over just exactly what kind of creature had possessed such a large tooth.

Figure 30-1. Reconstruction of the Cohoes mastodon. From John M. Clarke, *James Hall of Albany* (Albany, N.Y.: 1923).

Lord Cornbury, then the English governor of New York, pronounced it to be from a human giant. His interpretation was greatly influenced by his religion. Genesis 6:4 stated that there "were giants on the Earth in those days," and Cornbury thought this tooth had come from one of them. He sent people to search the site again, and soon a fair number of very decayed bone fragments were discovered to go with the tooth. The skeleton appeared to have been thirty feet long. A single thigh bone seemed to be seventeen feet long. Both estimates were no doubt wildly exaggerated.

There were others who thought that the remains must belong to some sort of animal or even a giant fish, but they could not tell just what kind. Congregationalist minister Edward Taylor was the first to suggest that the animal had been an elephant. But what kind of elephant? Back in the early 1700s nobody understood the concept of extinction, and so nobody was going to suggest that the Claverack Giant was anything other than a modern form of life, and there are, of course, no elephants in New York State.

Edward Taylor had other ideas, however. He thought that there might have been some thoroughly un-Christian giants. He cited Indian legends of ancient human giants. They were said to have been as tall as trees and they hunted bears. To profess pagan Indian myths, however, was not a good strategy in eighteenth-century New England, and Taylor's views were wisely kept to himself.

In 1706 some more mastodon remains were found, and these were shown to Massachusetts Governor Dudley. He thought that they were the eyeteeth of a human, and he added the notion that the specimen had been a victim of Noah's great deluge! His acquaintance, the influential Puritan Cotton Mather, of witch trial fame, embraced this notion enthusiastically. Despite his association with witch trials, Mather was something of a scientist. He sought to blend in some scientific rationalism in his theology. He saw the Claverack Giant as empirical proof of the Biblical story that there had been giants living in the pre-Deluge Earth.

It is true that there are some resemblances between the human molar and the mastodon's tooth, and back then little was known of fossil elephants, but Dudley's opinion is still just a little hard to understand.

Dudley's speculations had long faded into history by 1801, when a complete mastodon skeleton was found on Barber's farm in Orange County by Charles Wilson Peale. It was mounted and displayed at the Museum of the American Philosophical Society in Philadelphia, and was the first full skeleton of a prehistoric animal to ever be displayed to the public. Soon after, another skeleton was found nearby. It was also mounted, and sent off to Europe. Together they became ice age wonders that represented the might and grandeur of our new country.

In 1838, parts of another mastodon were found across the Hudson River in Greenville. This was, at the time, considered an important scientific discovery. New York's eminent paleontologist, James Hall, brought along the even more famous English geologist, Charles Lyell, to visit the site in 1841.

It wasn't long before another mastodon turned up in Greenville. In 1840 a partial skeleton was found somewhere along today's Rte. 32, maybe about a mile south of Greenville. This was within a "small swampy depression" on the farm of Charles Coonley. It is not clear exactly what was found here, and the bones seem to have been scattered, but it appears that a number of bones, perhaps from several elephants, came to light. This was long before careful excavations were done, and so this discovery will probably never be well understood.

We think that there are more of these to be found. These discoveries have been made mostly in the sediments of old glacial lakes and smaller kettle ponds, and there were a lot of those in the Hudson Valley (fig. 30-2). After the Ice Age ended, those lakes generally drained, and a large number of pools and wetlands were left behind (chapter twenty). If you travel around the Hudson Valley area, you will see these in abundance. Some of those wetlands have been dammed to create artificial ponds, but many of the pools and all of the swamps are natural. And any one of them, or all of them, might yield mastodons. But why are mastodons almost always found in wetlands?

Although it had long been thought that mastodons, because of the pointed cusps on their molars, may have preferred woody plants and shrubs for their diet, recent evidence has shown that their diets also included large amounts of grasses and aquatic plants. This may go a long way toward explaining why their remains have most frequently been recovered in wetlands of one type or another. Soon after the glaciers melted, and while early spruce and pine forests were beginning to establish themselves in the Hudson Valley, the numerous wetlands may have held the most abundant readily available food resources for mastodons. They also consumed shrubs, grasses, and small branches of trees, including the seed pods and cones. Elephants, after all, have very hearty appetites!

Back then, as now, winters brought ice to those ponds, and summers brought inviting spots filled with lush green vegetation and cool water. That gets us to those skeletons. The typical fossil mastodon is a young bull. Young males of most species have a reputation for being reckless, foolish, and always hungry, so it is so easy to imagine some among them wandering out onto the ice of a small pond and falling through it. Or

Figure 30-2. Cross-section of pond deposit showing the buried skeleton. Courtesy of New York State Museum.

they might be imagined as wading into the cool waters of summer for a nice snack and not being able to get back out of the pond. Elephants are bright animals, and they love water, but they don't rescue each other. Falling through the ice or wading a bit too far can be imagined as leading to entrapment in the thick muck. They were eventually preserved as fossil skeletons in the same pond sediments that had been their demise. An analogy to the La Brea tar pits of Los Angeles can be made. Out there, elephants got stuck in the tar; here, it seems, they became mired in the mud. Some skeletons, reportedly, have been found still standing.

As the millennia passed, the ponds turned into swamps and, in time, some of the skeletons were discovered. People dredge their swamps with the purpose of creating farmland or farm ponds, and encounter the sleeping giants of long ago. The drowned lands of Orange County fit the bill perfectly, and they have yielded a lot of post–ice age skeletons, many being mastodons.

What we find so interesting about all of this is that it tells us that there are still a lot of undiscovered mastodon skeletons throughout the region. Any swamp or pool of water is a potential hiding place for a mastodon, and there are many swamps and pools in the Hudson Valley. And, make no mistake about it, mastodons died in a lot of them. They are still there to be found.

Finding a mastodon is not easy; it requires luck, but it also requires that people know what to look for. Typically, somebody needs to make an excavation and work cuts into the old muds. Large bones are found, but too often people dismiss them as being of cows or horses. It's often the case that the significance of such discoveries is only understood when the tusks of the mastodon are found. It's pretty hard to confuse tusks with the remains of a cow.

So, what we are saying is that if you have seen some large bones in a local excavation, it is time to take another look. You might have found the bones of an ice age mastodon, and that should be made known. Good skeletons are worth a lot of money, too. We are told that a complete skeleton is worth about $50,000. We are also told that it costs a lot to properly excavate one—you guessed it, about $50,000.

31

Ice Age Ghosts

Geologists never know when they are about to take a trip into our distant past. It's just part of the job. Therefore, it wasn't a big surprise for us to begin one of those journeys recently when we were visiting Cedar Grove, Thomas Cole's home in the village of Catskill. As we turned around at the front door to admire the garden, we saw, and then took a good look at, a fine and very old honey locust tree.

The honey locust is only one among the many species of trees found throughout the Hudson Valley. It is certainly not the greatest of trees; there are bigger ones, and there are prettier ones. Nevertheless, there is something very special about this species. And once you have taken a good look at one, you will likely notice its distinction. Honey locusts are "armored" with very dangerous-looking thorns (fig. 31-1). Actually, "spike" is probably a better choice of words. These spikes can be three or four inches long, and they often occur in mean-looking clusters. The biggest clusters are found on the lower reaches of the tree's trunk. Up above, there are plenty more spikes strung out on the lower branches.

Anyone or anything careless enough to brush up against this tree will quickly find out what the spikes are for; they are vicious defense mechanisms. The lower branches hang down and seem to reach out with their spikes as if intending to do harm. Any browsing mammal will soon find out, and long remember, the dangers of trying to eat the foliage of this tree. The more intelligent the browser, the more effective the lesson.

Figure 31-1. A lower branch of a honey locust tree with its formidable spikes.

The defense functions are so obvious; but there is something else that is not immediately obvious. A puzzle emerges when you ponder these trees. That question is, "Who are these spikes defending against?" Your first guess might be white-tailed deer, especially if you are among those who have prized shrubbery in your yard. These are the dominant browsing animals of today's woods. But white-tailed deer would hardly be bothered by these spikes. Deer have slender snouts, and they have no problem finding plenty of space to pick at foliage between the spikes. No, locust trees have never much worried about deer.

But, if it is not the deer, then who is it? There are no other obvious browsers in today's woods. Those spikes also had to be aimed at something a lot bigger than a deer. And a lot taller too; they are very abundant for the first fifteen feet or so, and then thin out toward the middle and top of the tree. There is a real problem here; the fact is, there simply are no creatures in today's world that threaten our honey locusts. So, why do the trees go to all that trouble of growing those nasty long spikes?

Because there was one such browser long ago. Back at the end of the Ice Age, the Hudson Valley did have a great herbivore that might very well have pestered our honey locusts. And it was plenty large enough, too. That was the mastodon.

Modern elephants have a bad reputation for tearing up forests. They love to pull down limbs, and they are perfectly capable of stripping bark off the lower trunks of trees as well. In fact, elephants can virtually create their own habitat. They destroy so many trees that they break up the forests, creating lots of meadow in between the remaining patches of trees.

That rambunctious behavior, incidentally, also creates just exactly the right habitat for honey locusts. Locusts like to grow in broken forests, preferring to be right on the border between meadow and trees. So, it would seem that evolution may have cleverly adapted the locust to life with the mastodon. In turn, these great elephants may have created the habitat that was just right for locusts. And those large clusters of spikes protected the locusts from potential damage from the mastodons.

But there was more. The honey locust seed pods very likely appealed to the mastodons. Those seed pods hung just above the spikes; the elephants could just reach beyond the spikes, eat the pods, and then deposit the seeds elsewhere within their droppings.

All in all, the mastodons and honey locusts enjoyed a very fine symbiosis. But then, abruptly, it all ended. The mastodons went extinct about 11,000 years ago. The locusts lost the elephants that had helped them so much in reproduction. The locusts have survived to this day, but surely they are not as successful as they once were. Still, in the end, it is quite the concept to contemplate. These trees with their long spikes vigilantly wait for the elephants that will never ever come again. It is only the ghosts of mastodons that still haunt our forests.

32

Extinction

Like any couple, we have our occasional spats. But we are both scientists, so at times those heated moments are about scientific issues. We could never write a book like this together without having some disagreements, and so it has been with what is commonly called "the great megafauna extinction."

About the time that glaciers were finally melting and warmth was slowly returning to the Hudson Valley, many species of large animals that had long lived here were disappearing, never to return. Some species managed to hang on. Elk, bison, moose, caribou, and musk ox moved farther north as the climate warmed, while others like the peccary went south. They, or their close descendants, still thrive today.

But we lost forever some of the greatest, most wondrous, species that had once roamed the shores of the Hudson River. The mastodon and the mammoth, the dire wolf, the American horse, and the giant beaver are now known only from their fossil remains. Why did they disappear? What caused this great extinction? How about we give you some clues, and you can decide for yourself. But first, a bit of background is in order.

We have seen that, at the height of the last Ice Age, sea levels dropped dramatically and the more shallow sections of the ocean floor were exposed. At that time the strait between Siberia and Alaska was left high and dry. This area is known as the Bering land bridge, or Beringia. The land bridge provided a pathway for the migration of animals and people to and from Asia and North America. These people, called Paleo-Indians, were probably few in number at first, but they seemingly prospered in their new land. Their populations are envisioned as spreading like a wave across the continent and even down into South America, bringing with them efficient weapons and hunting skills.

Could the arrival and spread of the Paleo–Indians in the Americas have caused the extinctions of over 135 species at the end of the Ice Age? Since the 1960s, proponents of the "overkill" hypothesis think that is exactly what happened. They suggest that the Paleo–Indians, especially those called the "Clovis" culture (after their much-improved flint spear points [Fig. 32-1]), wandered across both continents, hunting with such efficiency and voracity that they managed to kill off most of the great herbivores. They so reduced the great herds of mega-herbivores that it became impossible for those populations to sustain themselves. This was quickly followed by the demise of the mega-carnivores—the American lions, saber-toothed cats, dire wolves, and short-faced bears. Those carnivores had lost most of the prey animals that they ate.

In its most extreme form this scenario becomes the "blitzkrieg" hypothesis proffered by Paul Martin, professor emeritus of the University of Arizona, who suggests that this hunting migration across the continents was nothing less than an outright killing spree. The Paleo-Indians migrated out of Alaska, and their populations expanded like a bubble all across the continent. All along the front of this expanding population bubble, they were killing off the megafaunas. This didn't stop until they made it to the east coast and then on down to the southern end of South America.

The only direct evidence to support either of these hypotheses are a very few scattered sites, mostly in the southwest, where spear points, and in some cases human remains, have been found in association with megafauna fossils. The close correspondence between the time of the appear-

Figure 32-1.
Clovis points, see fluted surface.

ance of Paleo-Indians and the time of extinction of the megafaunas has long been cited as evidence. And that may have happened on other continents. Australia witnessed a megafauna extinction of a similar sort soon after humans arrived there. Indirectly it is argued that, because the megafauna in all these locations had never before had a close association with people, these animals were naïve to the danger and therefore easy to hunt.

Some ethnobiologists and anthropologists who study the remains of early native sites argue against this hypothesis. Questions remain, mostly concerning the size and frequency of the migrations across the land bridge, how quickly their populations grew and spread across the continents, and whether those people killed indiscriminately. They also argue that some of the most probable target species of the Paleo–Indians, like bison and deer, were not driven to extinction. In fact, the populations of these species seem to have been little affected by human hunting pressure from the end of the Ice Age until the European expansion into the Americas. How had these species "escaped?"

Another group of scientists, led by Ross MacPhee of the American Museum of Natural History, also proposes the extinction of the megafauna by the same Paleo–Indians, but by different means. They hypothesize that these first humans arriving in the Americas brought with them, perhaps by way of their domesticated animals, a new but unspecified disease that the megafauna species had never before encountered. A "hyperdisease" that quickly wiped out so many, it made recovery of sustainable populations impossible and eventual extinction was the result. The debate is always lively. Opponents respond, saying that infectious diseases rarely have the ability to "jump" from species to species, especially with such lethal results for all. Rabies, a virus that infects all warm-blooded species, remains an exceptional example of how such a disease, if unchecked by modern medicine, might cause widespread extinctions.

Still other groups of scientists have hypothesized that the rapidly changing climate at the end of the Ice Age caused changes in vegetation types, which resulted in undue stress on herbivore populations that had been adapted to cold, arid conditions. Such changes could certainly have had a devastating effect on local populations, especially in areas where long periods of meltwater flooding from glaciers played a role. But there are problems with these climate change hypotheses. Could they account for the continent-wide extinctions that occurred? And hadn't some of these same species survived through other periods of glaciations?

The most recent hypothesis for the extinction of the megafauna blames an astronomical event or series of events. It argues that a comet or swarm of comets about 12,900 years ago caused severe fire and sudden climate changes that led to the mass extinction of not only the megafauna, but also quite likely most of the Clovis people. While there is some evidence of the deposition of a soot layer, called the "black mat"—indicating extensive fires—at different sites in North America during this time, it is doubtful that such an event impacted the entire Earth. And similar, but slightly younger, extinctions in South America would have had to have different causal mechanisms.

In the end, there is certainly no shortage of hypotheses seeking to explain the disappearance of large mammals at the end of the Ice Age. Are any of these the real reason why the megafauna disappeared? Could some combination of them explain why there are no mastodons left in the Hudson Valley today? Well, the two of us continue to devote a fair amount of our time to the debate.

33

The Dunes of Pine Bush

A large urban park is an unlikely phenomenon, when you think about it. It's not like there aren't any—Central Park in New York City, and the Fens in Boston come to mind—but land is at a premium in a city, and there is only so much to spare. Albany, however, has something even more unusual—an urban nature preserve. Those are very rare indeed. This one is called "the Pine Bush," and it is located on the western side of the city. Actually, it is composed of a number of parcels of land that collectively are called the Pine Bush. As the name implies, there are a lot of pines and a lot of bushes, but it is not the living ecology that is our focus here. It is, of course, the curious geological history of this part of Albany.

To understand the Pine Bush we have to take another of our many journeys back about 15,000 years, back to what was, in these parts, a time when the Ice Age was coming to an end. The Albany area would have been remarkably different from today. It was covered with an ice age lake (fig. 27-2). A mind's eye aviator, flying above this part of the Hudson Valley, would have looked north and observed a great glacier extending from the west to the east, clear across the valley. This was the

Hudson Valley glacier with which we have become so familiar. It was a dynamic time for this ice. The climate was balanced between being just cold enough to support a glacier and just warm enough to melt one. The ice was likely to be actively moving southward in the Clifton Park area, but at Albany it can be envisioned as already melted away.

To the west, the Mohawk River was flowing actively, and it carried dirty water to the Schenectady area. Here, there was Glacial Lake Albany, extending off to the southeast along the flank of the melting Hudson Valley glacier. Farther south, most of the Hudson Valley was flooded. Not surprisingly, the dirty waters of the Mohawk produced a large delta. Here, it's a very large heap of sediment, mostly sand. In fact, almost all of Rotterdam and Schenectady are perched upon an old delta deposit (fig. 27-2).

The climatic changes were slow but persistent; as the climate warmed, the ice had melted and the glacier had retreated. Gradually, the ice had been vacating the Albany area. It was replaced by the waters of a growing Lake Albany. There was plenty of sediment being eroded from the surrounding highlands, as there were not yet many plants to slow down rates of erosion. So it was that great quantities of sediment accumulated at the bottom of this sizable lake.

Once again we must make a point about these glacial lakes. In this sort of science, the hardest thing to see is that which is not there. This is the case with the glacial lake at Albany. You have probably never noticed it, but Albany is not a San Francisco; there are very few hills. Almost all of Albany is, in fact, as flat as the proverbial pancake. The next time you are there, look around and you will see this. Albany is built upon the floor of Glacial Lake Albany, and we have seen, up and down the Hudson Valley, that this is very flat indeed.

But, to the west, the flatness disappears and a series of low, rolling sandy hills are found; that is the Pine Bush. Naturally, people prefer to develop flat landscapes, and they put off building on hilly areas because it is just harder to do. Possibly, on that account, the Pine Bush area was slow to be developed, and that saved it long enough for people to be able to make it into a nature preserve. One at a time, bits of land were added to the preserve, and today it is a series of mostly unconnected parcels linked by one administration and also by one geological heritage.

The Pine Bush hills are fossil sand dunes. They date back to the time we are speaking of, when the Ice Age was ending. As the Hudson Valley ice retreated northward, the waters of Lake Albany followed. But this great ice age lake was doomed. As the weight of the ice was removed from the area, the crust, which had been depressed, actually rose (isostatically, see chapter ten). As the landscape rose, the waters of Lake Albany were dumped unceremoniously down the Hudson River and into the Atlantic Ocean. Gradually the lake waters became shallow, and slowly the lake bottom became dry land. When this happened at the Rotterdam/Schenectady delta, large amounts of sand became exposed and dried out.

The weather patterns back then, like today, blew out of the west. Winds howled across the old delta sediments, and much of the sand blew to the east. Windblown sand is very good at forming into sand dunes, and that's exactly what happened at Pine Bush. It's quite something to imagine. For a substantial period of time, this part of Albany was a cold-climate desert. Large, treeless dunes migrated across the countryside, driven by the wind. Toss in a few camels, and it certainly gives you a very different impression of Albany. But there were no camels.

With time and improving climate, a succession of plants colonized the Pine Bush. Eventually the penetrating roots of grasses and shrubs came to stabilize the dunes. Still later in time, pine forests appeared, and the Pine Bush came to resemble its modern form. All this brought dune migration to a halt and left the many dunes essentially frozen in their tracks. That makes the Pine Bush something of a fossil landscape. It may be a modern plant ecosystem, but beneath this forest is an ice age landscape.

And that fossil landscape is something that you can enjoy anytime. The Pine Bush preserve has a system of hiking trails that are open all year. Most of the lengths of these trails are across flat terrain. That makes for nice, easy walking, but you should appreciate the flat for what it is; the flat represents, geologically, the sands that were deposited upon the bottom of the old ice age lake that was once here. Stop and look up into the sky, and imagine the water that once rose above.

In your mind's eye you can see the sunlight passing through the waves above and penetrating to the sandy floor of the lake all around. Up there, the wind is blowing out of the west, and waves are passing by. They catch the sparkle of the sunlight as they pass. Occasionally a small cake of ice floats by; sometimes there are large blocks of ice to be seen. They bob up and down in the waves. There are no fish and no water birds; after all, it is still the Ice Age.

September 25, 2009—We began our hike down the Great Dune Trail, heading west from the Willow Street trailhead. Soon we were climbing up and over the slopes of an old sand dune. We found that one of the older trails had been worn into a deep rut. After passing over the crest, we found the slope beyond to be even steeper. This was the "**slip slope**" (fig. 33-1) of the old dune. Long ago the wind had blown sand across the crest and down this steep slope. Soon we found ourselves on the yellow trail. We crossed another dune, and then the trail turned east. The trail soon found its way to the base of the slip slope of the "Great Dune." This is a wonderful large old fossil dune. It stretched almost a mile off to the

Figure 33-1. A dune slip slope.

Figure 33-2. Swale ponds.

east. It has a fine slip slope facing to the southeast and a gentler slope on the other side. All along the base of the dune, we found small ponds and wetlands (fig. 33-2). These were the remnants of what are sometimes called **swale ponds**. Back at the end of the Ice Age, when the dunes were just stabilizing, a series of ponds formed in the low troughs in front of the dunes. In the thousands of years that have followed, most of these ponds have filled in with sediment and now they are just wetlands, tiny versions of the "drowned lands." A few remain as small pools.

We continued on, almost to the eastern end of the great dune. There the slip slope rose steeply to its crest. We climbed up to the top and gazed beyond it ...

September 25th, 12,990 BP—We stood on the crest of the Great Dune. There were no trees here, and all around was the blinding white of quartz sand. A powerful wind was blowing in from behind us, sometimes gusting to very high speeds. We felt sand grains stinging the backs of our necks and ankles. We looked behind and saw sand grains bouncing up the slope. The wind was sculpting them into small ripples. In the distance we saw more white dunes. The sky was clear and the sun was bright. Our eyes squinted to avoid the glare.

We turned and looked downslope to the southeast. At the bottom of the dune was a long pond. There were no plants around it, just more bare white sand. The water was deep, but there were no signs of any living creatures within. Then, off to the far left, we saw tracks leading to and away from the pond. They were the large footprints of an elephant. "Who are you?" we wondered. "You can't be a mastodon. There shouldn't be any here; they like to stay in the forests, and there is no forest here." Could it have been a mammoth? Maybe, but there are few mammoths to be found in this part of the state. We pondered the question, but we would never find out the answer to this mystery; our stay in this moment of time would be too brief.

34

Bad Day on Wall Street

Some days are worse than others. We would like to describe one of the worst ever for the whole of the Hudson Valley, and for much of the Champlain Valley to boot. This was an environmental disaster about as bad as can be imagined. Like so many others that we worry about nowadays, this one involved global warming. But this was warming on a scale that was much more extensive than any of us will ever see … we hope. Let's get on with the story.

We have already set the stage for this. The last ice age began to come to an end about 21,000 years ago. A great continental ice sheet had advanced as far as Long Island and, as the climate warmed, it began to retreat. Researchers at Woods Hole Oceanographic Institute, among others, have recently found that, by 13,400 years ago, the ice had melted back to the northern tip of the Adirondacks. That set up a very perilous situation. Just to the west was an enormous ice age lake. It was called Glacial Lake Iroquois (fig. 34-1). It was located where Lake Ontario is, but it was about three times bigger. Take a good look at that map and appreciate just how large this lake was.

The waters of the lake were trapped behind dams composed of

the Adirondacks to the south and east, and the ice sheet itself farther to the north (fig. 34-1). But that dam was about to break. This was a time of *serious* global warming. Glacial meltwater already flowed south through two large lakes. One, Glacial Lake Vermont, filled the Champlain Lowland. The other, Glacial Lake Albany, we have seen filling much of the Hudson Valley. So, most of the region was already underwater, but it was going to get much worse.

Back north, the retreating ice was about to leave behind just a tiny little hole in the dam. As the ice sheet melted away, it retreated from the northern tip of the Adirondacks, and water rushed through what was at first just a very small gap. The opening quickly became a very large one, and then an icy tsunami cascaded down through Lake Vermont and on to Lake Albany.

To call this a catastrophic flood is an enormous understatement. We have to give some more numbers that won't mean much, but will communicate just some of the magnitude of this event. In a very short period, Glacial Lake Iroquois lost about 160 cubic miles of its water. Think about how much water that involved. It was about 120 million gallons of water per second — for about 88 days! All of that water was funneled into those two relatively narrow lowlands. That spelled major trouble.

This gets us back to the Hudson Valley. For those 88 days the water kept coming relentlessly, flowing south down the Hudson. Lake Albany must have swelled upwards and turned into a roiling, churning, foaming chaos of broken ice and frigid ice water. The power and rage of that flow are difficult to imagine. None of us will ever see the like of this ... we hope.

What we are portraying is a rising of the Lake Albany waters to levels far above what had been seen—but not for long. Down the Hudson Valley, near today's Verrazano Narrows Bridge, there was the great earthen dam, the old terminal moraine. Floodwaters soon backed up behind that moraine dam, rose up over it, and then began eroding into it. First a small notch was cut, and then waters poured through. Soon the flow cut a great gap in the dam, all the way down to bedrock.

With the dam breached, the water of Lake Albany rapidly poured out onto the still-exposed continental shelf, over the edge into the Hudson Canyon, and downwards on to the Atlantic. Throughout the Hudson

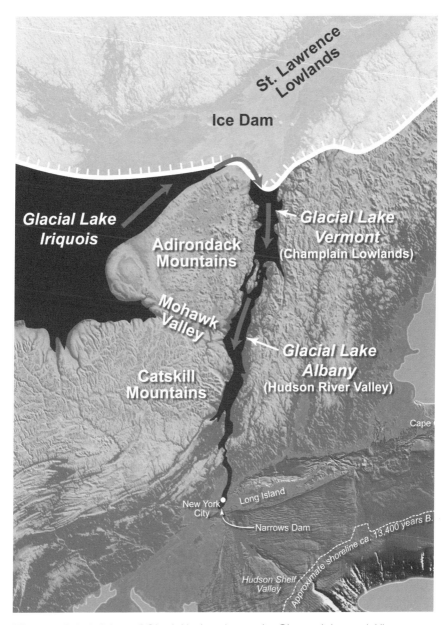

Figure 34-1. Map of Glacial Lakes Iroquois, Champlain, and Albany. Illustration by Jack Cook of the Woods Hole Oceanographic Institute.

Valley that meant a sudden drop in water levels. Soon the lake waters shrank down to a narrower, but very erosive, flow, helping to renew and deepen the ancient river channel known as the Hudson Shelf Valley. This quite possibly helped cut a deep Hudson River channel. When all this was done and things settled down, the Hudson River may well have looked a lot more like its modern self. Something similar to today's river flowed by.

But, before then, that mighty erosive flow must have taken quite a toll upon our local geography. We wish we could point out some features that were shaped at that time but, for now, we can't. We can tell you that there are reported to be boulders the size of small cars that have been found offshore of New York City, and these are thought to have been carried out to sea by this tsunami. It takes a strong current to move objects like that, and it almost frightens us to just imagine how strong. But, some days are worse than others.

35

Never Again?

About 700 million years ago, the relentless motion of the Earth's crust brought the drifting of most of the world's continents onto the equator. That's unusual, and it is certainly something that happened very rarely in our planet's history. It was just something of a statistical oddity, but the results were dramatic. Continents tend to be far cloudier than oceans, and the clouds, being white, reflect sunlight back into space. If enough sunlight bounces back into space, the result is a cooler Earth, and that's what happened.

Soon the polar seas began to freeze and glaciers formed on adjacent continents. Glaciers also are white, and so more sunlight was reflected back into space and the world became even colder. And, as it was colder, the world saw more glaciers form and advance closer to the equator.

All this completely disrupted the planet's climatic equilibrium, and in short order there was a **runaway glaciation**. Glacial ice advanced from the poles all the way to the equator, and the great blue marble that had been the planet Earth became white. This is called the "snowball Earth" hypothesis. You can forgive us if we call this one of the most *chilling* discoveries in the history of geology.

Today most of our equatorial seas are open water, and a runaway ice age is hardly likely to happen again. But it is fair to ask if the ice ages are truly over, or are there prospects that the Earth may descend into a deep freeze once again. Will glaciers ever again come down the Hudson Valley, and if so, when?

We geologists search the record of Earth's history to find the answers to such questions, and answers can be found. There have been several other ice ages in our history. An early large glaciation occurred a bit more than two billion years ago. Then there was another one between 600 and 800 million years ago. That is the one we just described at the start of this chapter; it was perhaps the most extensive, the "snowball Earth" event. Lesser glaciations occurred in Africa and South America between 460 and 260 million years ago. There was, at that time, a great southern supercontinent, called Gondwanaland, and it straddled the South Pole. All of Antarctica and India, and much of South America, Africa and Australia were linked and lay upon the South Pole. It was, of course, a cold climate, and glaciers formed and extended all across these lands. The next and most recent events were the Pleistocene glaciations, some of which have been the topic of this book.

The long and the short of it is that ice ages are not unique events; they have happened sporadically throughout the history of our planet. That means there is no reason to suppose that they will not happen again. We have, earlier, speculated about this (chapter two) and suggested that a return of the ice might occur some 50,000 years from now.

We won't spend a lot of time defending this notion; it has no immediate importance to any of us living today. But it seems inevitable that the glaciers *will* come again sometime in the future. That means that all of the imagery we have devoted to describing past glacial events throughout this book apply equally to the future as well as to the past. Take another look at our chapters one and twelve. After that, find some of our numerous other journeys into the past. Then look into the future; it will all happen again.

Bibliographic Essay

Introduction

Typically books of this sort have extensive chapter notes and references. A reader who is interested in pursuing a topic mentioned in the book might search through those chapter notes and be directed toward further research materials. You may be interested in learning more about some topic in our book, and that is why we have added this section as a quick reference guide. But, times have changed and facts are facts; if you want to pursue some topic, do what everybody does nowadays—Google it! The quality of information available on any particular Web site can vary considerably from poor to excellent, and even Wikipedia entries have been controversial, but we have found online sources to be generally quite acceptable, especially for the lay reader, so it is a good place to begin. As it happens, a lot of this book, especially chapters 12–24, is biographic. We describe our adventures exploring the Hudson Valley, and this does not require serious referencing. We recognize that the great majority of our readers are not professionals and are not likely to want to do serious research. Still, we would like to be of some assistance for those who do want to learn more, hence the following.

Terms: We use a lot of technical terms throughout, and they are highlighted in bold type wherever introduced. We hope that we have defined them all well enough, but you might want to look at some references. We recommend a couple of good geomorphology textbooks. The first is Don Easterbrook's *Surface Processes and Landforms* (New York: MacMillan, 1993). Then there is Arthur Bloom's *Geomorphology* (Englewood Cliffs: Prentice Hall, 1991) and Ritter, Kochel and Miller's *Process Geomorphology* (Boston: McGraw-Hill, 2002).

General References: We recommend two broad-ranging references. One is the *Surficial Geological Map of New York*, compiled and edited by Cadwell and Dineen (Albany, N.Y.: New York State Museum, 1987). It comes in four sheets that cover the whole state; two are pertinent to the Hudson Valley. We carried these up and down the valley in our explorations. The other is *The Wisconsinan Stage of the First District, Eastern New York*, Donald Cadwell, editor, New York State Museum Bulletin 455, 1986. This, too, was an indispensible reference in our work, with many articles about the region. It's getting a little old, but it is still valuable.

Causes of the Ice Age: This is a very dangerous topic. There is a lot of disagreement among members of the profession. Different authors take different approaches to the problem, and this is not a settled issue. Late at night in geology bars, this topic has been debated for time immemorial. We recommend any recent historical geology text and have no strong preferences. Some we can suggest include Prothero and Dott's *Evolution of the Earth,* 8th edition (Boston: McGraw-Hill, 2011). You might also try Steven Stanley's *Earth System History,* 2nd edition (New York: Freeman, 2004). Another book is William Ruddiman's *Earth's Climate—Past and Present* (New York: Freeman, 2001). Typically, a good historical geology text will describe topics we have covered, such as the greenhouse effect, cycles of glaciation, and Milankovitch Cycles. But, of course, all these can be found through Internet search engines like Google. Don't be surprised when you find that opinions on all these terms and topics vary considerably. That's the nature of this kind of science.

The Advance of the Ice: Again, a good historical geology textbook should contain something on this topic. You might want to read more about the Shawangunk Mountains, and we recommend Jeffrey Perls's *Shawangunk Trail Companion* (New Paltz, N.Y.: Countryman Press, 2003). Another one we liked was Jack Fagan's *Scenes and Walks in the Northern Shawangunks* (New York: NY/NJ Trail Conference, 2006). A reference more focused on the trails was Edward Henry's *Gunks Trails: a Ranger's Guide to the Shawangunk Mountains* (Delmar, N.Y.: Black Dome Press, 2003).

The Hudson Canyon: Here, let's just send you to a Web site that has some very good graphics: http://www.whoi.edu/page.do?pid=9779&tid=3622&cid=2078

Retreats of the Ice: This topic has been well covered by publications of the New York State Museum. We like New York State Museum Bulletin 497, 2003, especially the article by John Ridge. See his figures 3.7 and 3.8, New York State Museum Bulletin 455, edited by Donald Cadwell, presents a number of retreat ice margins. We used several of these maps in our explorations.

The Drowned Lands: Once again, another Web site is a good place to look: http://www.pineislandny.com/history.htm

Lake Albany: We were disappointed to see how hard it is to find a good map of Lake Albany. The map we used is from New York State Museum Bulletin 497. Our online search was not very fruitful.

The Draining of Lake Albany past New York City: This is another of those highly debated issues in the history of Hudson Valley glaciation. We will cite one technical Web site in support of the Arthur Kill hypothesis: http://www.geo.hunter.cuny.edu/bight/raritan.html

Hudson Valley Landscape Architecture: There is an abundance of literature devoted to landscape architecture and its nineteenth-century development, much of which occurred in the Hudson Valley. We especially like Robert Toole's *Landscape Gardens on the Hudson: A History* (Delmar, N.Y.: Black Dome Press, 2010). See his introductory chapter, along with chapters about Montgomery Place, Hyde Park and Wilderstein. See also one good Web site: http://www.hudsonrivervalley.org/themes/landscape.html/

Hudson Valley Faulting: The only reference we have ever seen on this topic is J.B. Woodworth's "Postglacial Faults of Eastern New York" in New York State Museum Bulletin 107, 1907. When we went and looked, we could not find any of the faults that Woodworth reported.

Slumps: This has been an issue that Robert Titus has frequently covered in a number of his newspaper and magazine columns. See "Landslides on the Highway?," *Kaatskill Life*, Vol. 26, no. 1, spring 2011. See also this Web site: http://www.bgs.ac.uk/discoveringGeology/hazards/landslides/rotationalSlides.html

North Atlantic Oscillation: This is a subject covered by difficult technical writings. We recommend one Web site: http://www.ldeo.columbia.edu/res/pi/NAO/. Its graphics are, unfortunately, a bit out of date.

Ice Age Paleontology: Mastodons have been covered most extensively under this broad heading. An old classic is *Mammoths and Mastodon Ice Age Elephants of New York* by Judith Drumm, New York State Museum, Educational Leaflet 13, 1963. An important reference is *American Monster*, Paul Semonin (New York: New York University Press, 2000).

Honey Locusts: We recommend Connie Barlow's *The Ghosts of Evolution* (New York: Basic Books, 2000). The book tells the story of the honey locust along with other similar tales of late– and post–ice age evolution.

See also one Web site: http://outwalkingthedog.wordpress.com/2010/03/29/mastodons-in-manhattan-how-the-honeylocust-tree-got-its-spikes/

Megafauna Extinction: The best-known author concerning this topic is Paul Martin. See his *Twilight of the Mammoths: Ice Age Extinction and the Rewilding of America* (Berkeley: University of California, 2005). Peter Ward is another fine popular science writer who has dealt with the subject. See his *The Call of Distant Mammoths: Why the Ice Age Mammals Disappeared* (New York: Springer-Verlag, 1997). We also recommend Sharon Levy's *Once and Future Giants: What Ice Age Extinctions Tell Us about the Fate of Earth's Largest Animals* (Oxford/New York: Oxford University Press, 2011). Let's also add a few Web sites, covering some of the latest research:

http://www.eurekalert.org/pub_releases/2011-11/uoc-utc103111.php

http://www.nature.com/nature/journal/v479/n7373/full/nature10574.html

http://news.nationalgeographic.com/news/2001/11/1112_overkill.html

Pine Bush: An extensive description of the Pine Bush can be found in Jeffrey Barnes's *Natural History of the Pine Bush* in New York State Museum Bulletin 502, 2003. A far more technical discussion can be found in Robert Dineen's *The Geology of the Pine Bush Aquifer, North Central Albany County, New York*, New York State Museum Bulletin 449, 1982. Good trail maps of the Pine Bush Preserve are available at the site.

The Great Flood: See the following Web site, especially for its graphics: http://www.whoi.edu/oceanus/viewArticle.do?id=5078

Snowball Earth Hypothesis: This is a relatively new concept only found in the more recent historical geology textbooks. See Prothero and

Dott's *Evolution of the Earth*, 8th edition (Boston: McGraw-Hill, 2011) for a good discussion. We recommend one Web site: http://en.wikipedia.org/wiki/Snowball_Earth

Photograph of Johanna and Robert Titus by Linda Post.

About the Authors

Robert Titus, PhD, is a paleontologist by training who has done considerable professional research on the fossils of upstate New York. He teaches in the Geology Department at Hartwick College. His previous books—*The Catskills: A Geological Guide* (3rd edition, 2004), *The Catskills in the Ice Age* (revised edition, 2003), *The Other Side of Time: Essays by "The Catskill Geologist"* (2007)—were published by Purple Mountain Press. Johanna Titus, MS, has a degree in molecular biology. She teaches in the Allied Health and Biological Sciences Department at SUNY Dutchess. Robert and Johanna write regular columns for *Kaatskill Life* magazine, the Register Star newspaper chain and the *Woodstock Times*.

Index

A
Acadian Mountains, 100
Agawamuck Creek, 133–136
albedo, 13, 14
alluvial fan, 103–108
Alpine glacier, 107
Arthur Kill, 62, 77

B
Bash Bish Brook, 122–123
Bash Bish Falls, 121–123
Berkshire Mountains, 34, 52, 62, 125
bottomset, 68
Boulder Rock, 98
braided streams, 1, 56, 124

C
Cairo Round Top, 96, 97
Catskill Delta, 100
Catskill Front, 29–34
Cedar Grove, 179
Central Park, 28
chattermarks, 22–23
Church Hill, 146–148
Claverack Giant, 175
cloven deltas, 69
Clovis culture, 184, 186, 187
Columbia Land Conservancy, 125, 133
crescent marks, 22–23
Crum Elbow Creek, 167

D
dead ice, 79
deranged drainage, 131–132
deltas, 67–74
Doove Kill, 117–119
drowned lands, 57–60, 126–128, 174, 178
Drowned Land Swamp Conservation Area, 125
drumlins, 137–140, 141–144
Dutcher Pass, 91, 94

E
Earth's eccentricity, 15–16, 17
Earth's obliquity, 16
Elizaville Delta, 71–74
Emmons, E., 157
Esopus Bend Nature Preserve, 64

F
Fawn's Leap, 99, 105–108
foreset, 68

G
Gertrude's Nose, 26
glacial epochs, 9, 17
glacial erratics, 26
Glacial Lake Albany, 58, 61–66, 67, 69, 77, 78, 140, 150, 163, 166, 188, 189, 194, 195
Glacial Lake Connecticut, 49

Glacial Lake Cooper, 113
Glacial Lake Flushing, 49, 77
Glacial Lake Hackensack, 49, 77
Glacial Lake Iroquois, 46, 193–195
Glacial Lake Kiskatom, 110–114
Glacial Lake Passaic, 49, 77
Glacial Lake Wallkill, 58
Glacial Lake Woodstock, 113
Greenhouse Earth, 8
Greenland Center, 13
ground moraine, 24
Gulf Stream, 12–13, 14

H

hanging deltas, 69
High Falls (Philmont), 133–136
honey locusts, 179–182
Hudson Canyon, 41–46
Hudson River School of art, 145
Hudson Shelf Valley, 42
Hudson Valley Glacier, 21, 29–34, 41, 42, 48, 50, 51, 54, 56, 79, 83, 89, 188
Hyde Park Delta, 151, 167

I

ice ages, 7–18
ice cored delta, 71
Icehouse Earth, 8–9, 11
ice margin, 52
ice sheet, 5, 17
insolation, 15–17
interglacial epochs, 9, 17
isostatic rebound, 62, 150, 167, 189

Isthmus of Panama, 12

J

joints, 30–32

K

Kaaterskill Clove, 99–108, 112
Kaaterskill Falls, 136
kames, 38, 54, 131
Keewatin Center, 13
kettle ponds, 38, 40, 176
kettles, 38, 54
Kiskatom Flats, 110–114

L

Labrador Center, 13, 35, 36, 40
Lake Taghkanic, 115–117
landscape architecture, 145–148
Laurentide ice sheet, 5, 29, 32, 33, 35, 42, 52, 62, 77–78, 83
Long Island, 34–40, 52, 54, 75

M

Martin, Paul, 184
mastodons, 173–178, 181
megafauna extinction, 183–186
Milankovitch cycles, 15
Millbrook Mountain, 20–28
Minnewaska Park, 20, 21–28
Montgomery Place, 149, 151–152
moraine/outwash complex, 39, 54
Mount Marion Delta, 74
Mount Merino, 146–147

N

North Atlantic Oscillation (NAO), 165–166
North Lake, 28, 32
North/South Lake State Park, 28, 89–98

O

Olana, 146–148
outwash plain, 39, 54, 55, 56, 87, 124
overkill hypothesis, 184

P

Palenville fan, 102–105
paleoforms, 94
Paleo-Indians, 184
Palmaghatt Kill, 27
Patterson's Pellet, 26
Pellets Island Moraine, 60
Philmont, 133
Pine Bush, 187–192
Plattekill Clove, 32, 33, 83–88, 105
Plattekill Creek, 74
Pleistocene, 9, 13, 198
plucking, 25, 32
plunge pool, 94
post–ice age faults, 155–158
precession, 16, 17

R

readvances, 53
recessional moraine, 40
Red Chasm, 99

Red Hook moraine, 54, 55
Rip's Rock Ledge, 89
rock drumlins, 146
Roeliff Jansen Kill, 65, 121–123, 125–128
Ronkonkoma moraine, 37
rotational slumps, 161
runaway glaciation, 197

S

Scandinavian Center, 13
Schenectady Delta, 162, 188
Schoharie Creek Valley, 86–88
Shawangunk Mountains, 20–28
silent earthquake, 159–162
"Sleeping Giant," 20
slumps, 168–172
snowball Earth, 197
South Mountain, 98
spillways, 90–98, 116, 119
Springwood, 149, 168, 170
stillstands, 52
striations, 22, 28, 32, 33
swale ponds, 191

T

Taconic Mountains, 34, 52, 56, 62
Taconic State Park, 121
Taghkanic State Park, 115–117
Tappan Zee, 4, 78
terminal moraine, 37, 77
topset, 67
tundra, 2
turbidity current, 44

V

Vanderbilt Mansion, 69, 70, 149, 151, 168–172
varves, 63–64

W

Wallkill Valley, 57
Wall of Manitou, 28, 29
wetlands, 129–132
white-water streams, 58
Wilderstein, 152–154
Wilson cycles, 9–14, 17
Wisconsin Stage, 9
Woodstock Land Conservancy, 133
Woodworth, J.B., 157

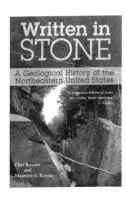

WRITTEN IN STONE
A Geological History of the Northeastern United States
by Chet Raymo & Maureen E. Raymo
(trade paperback, 6 x 9, 192 pages, illustrations, charts, maps,
ISBN 9781883789275, $16.95)

"Compresses billions of years into a slim, lively narrative."
Yankee magazine

In vivid, nontechnical prose, two university professors—a father and daughter team—trace the geologic changes in the American northeast since the continent perched on the equator and dinosaurs were young. Continents collide, oceans disappear, mountain ranges rise and fall, and mass extinctions decimate entire species. *Written in Stone* is an indispensable reference that "will captivate the interested layperson and refresh the professional" (*Science News Books*).

ABOUT THE AUTHORS

Chet Raymo is Professor Emeritus at Stonehill College in North Easton, Massachusetts. He is the author of more than a dozen books on science and nature. His work has been widely anthologized, including in the *Norton Book of Nature Writing* and annual editions of *Best American Science and Nature Writing*. He is a winner of a 1998 Lannan Literary Award for his nonfiction work, and the subject of a biographical essay in Scribner's two-volume *American Nature Writers*. For over twenty years he explored the relationships between science, nature and the humanities as columnist for the *Boston Globe*.

Maureen E. Raymo, Chet's daughter, is a paleoclimatologist/marine geologist with the Lamont-Doherty Earth Observatory of Columbia University. She has authored numerous scientific publications on the topic of Earth's climate and how it has changed in the past.

AVAILABLE FROM
Black Dome Press, 1-800-513-9013, blackdomepress.com